COMPUTER CODING

CODING

FOR KIDS

computer coding

FOR KIDS

A UNIQUE STEP-BY-STEP VISUAL GUIDE,
FROM BINARY CODE TO BUILDING GAMES

DK UK

Project editor Ben Ffrancon Davies
Editor Sam Priddy
Designer Fiona Macdonald
Additional editors Sam Atkinson, Lizzie Davey, Daniel Mills, Ben Morgan
Additional designer Simon Murrell
Consultant editor Craig Steele
Managing editors Lisa Gillespie, Paula Regan
Managing art editor Owen Peyton Jones
Producers, pre-production Ben Marcus, Jacqueline Street
Senior producers Meskerem Berhane, Mary Slater
Jacket editor Emma Dawson
Jacket designer Surabhi Wadhwa
Jacket design development manager Sophia MTT
Publishers Sarah Larter, Andrew Macintyre
Art director Karen Self
Design director Phil Ormerod
Associate publishing director Liz Wheeler
Publishing director Jonathan Metcalf

DK INDIA

Senior editor Suefa Lee
Senior art editor Devika Dwarkadas
Project editor Tina Jindal
Project art editor Sanjay Chauhan
Editor Neha Pande
Art editors Rabia Ahmad, Simar Dhamija, Sonakshi Singh, Shreya Anand Virmani
Assistant art editor Vanya Mittal
Jacket designers Priyanka Bansal, Suhita Dharamjit
Jackets editorial coordinator Priyanka Sharma
Managing jackets editor Saloni Singh
DTP designers Jaypal Singh Chauhan, Sachin Gupta, Rakesh Kumar
Senior DTP designer Harish Aggarwal
Managing editor Rohan Sinha
Managing art editor Sudakshina Basu
Pre-production manager Balwant Singh

This edition published in 2019
First published in Great Britain in 2014 by
Dorling Kindersley Limited
DK, One Embassy Gardens, 8 Viaduct Gardens, London, SW11 7BW

The authorised representative in the EEA is Dorling Kindersley Verlag
GmbH. Arnulfstr. 124, 80636 Munich, Germany

A CIP catalogue record for this book is available from the British Library.
ISBN: 978-0-2413-1773-0

Printed and bound in China

For the curious
www.dk.com

This book was made with Forest Stewardship
Council™ certified paper – one small step
in DK's commitment to a sustainable future.
For more information go to
www.dk.com/our-green-pledge

CAROL VORDERMAN MA(CANTAB), MBE is one of Britain's best-loved TV presenters and is renowned for her mathematical skills. She has hosted numerous TV shows on science and technology, from *Tomorrow's World* to *How 2*, and was co-host of Channel 4's *Countdown* for 26 years. A Cambridge University engineering graduate, she has a passion for communicating science and technology, and has a keen interest in coding.

DR JON WOODCOCK MA(OXON) has a degree in Physics from the University of Oxford and a PhD in Computational Astrophysics from the University of London. He started coding at the age of eight and has programmed all kinds of computers from single-chip microcontrollers to world-class supercomputers. His many projects include giant space simulations, research in high-tech companies, and intelligent robots made from junk. Jon has a passion for science and technology education, giving talks on space and running computer programming clubs in schools. He has worked on numerous science and technology books as a contributor and consultant.

CRAIG STEELE is a specialist in Computing Science education who helps people develop digital skills in a fun and creative environment. He is a founder of CoderDojo in Scotland, which runs free coding clubs for young people. Craig has run digital workshops with the Raspberry Pi Foundation, Glasgow Science Centre, Glasgow School of Art, BAFTA, and the BBC micro:bit project. Craig's first computer was a ZX Spectrum.

SEAN McMANUS learned to program when he was nine. His first programming language was Logo. Today he is an expert technology author and journalist. His other books include *Scratch Programming in Easy Steps*, *Web Design in Easy Steps*, and *Raspberry Pi For Dummies*. Visit his website at www.sean.co.uk for Scratch games and tutorials.

CLAIRE QUIGLEY studied Computing Science at Glasgow University, where she obtained a BSc and PhD. She has worked in the Computer Laboratory at Cambridge University and Glasgow Science Centre, and is currently working on a project to develop a music and technology resource for primary schools in Edinburgh. She is a mentor at CoderDojo Scotland.

DANIEL McCAFFERTY holds a degree in Computer Science from the University of Strathclyde. He has worked as a software engineer for companies big and small in industries from banking to broadcasting. Daniel lives in Glasgow with his wife and two children, and when not teaching young people to code, he enjoys cycling and spending time with family.

Contents

4 INSIDE COMPUTERS

5 PROGRAMMING IN THE REAL WORLD

Find out more at:
www.dk.com/computercoding

Foreword

Just a few years ago, computer coding seemed like a mysterious skill that could only be practised by specialists. To many people, the idea that coding could be fun was a strange one. But then the world changed. In the space of a few years, the Internet, email, social networks, smartphones, and apps hit us like a tornado, transforming the way we live.

Computers are a huge part of life that we all now take for granted. Instead of calling someone on the phone, we send a text message or use social media. From shopping and entertainment to news and games, we guzzle on everything computers have to offer. But we can do more than just use this technology, we can create it. If we can learn to code, we can make our own digital masterpieces.

Everything computers do is controlled by lines of code that someone has typed out on a keyboard. It might look like a foreign language, but it's a language anybody can pick up quite quickly. Many would argue that coding has become one of the most important skills you can learn in the 21st century.

Learning to code is tremendous fun as you can get instant results, no matter how much more you have to learn. In fact, it's such fun creating games and programs that it feels effortless once you're hooked. It's also creative – perhaps the first science that combines art, logic, storytelling, and business.

Not only that, coding is a fantastic skill for life. It strengthens logical thinking and problem-solving skills – vital in many different areas of life, from science and engineering to medicine and law. The number of jobs that require coding is set to increase dramatically in the future, and there's already a shortage of good coders. Learn to code, and the digital world is yours for the taking!

CAROL VORDERMAN

How this book works

This book introduces all the essential concepts needed to understand computer coding. Fun projects throughout put these ideas into practice. Everything is broken down into small chunks so that it's easy to follow and understand.

Pixel people give hints and tips along the way

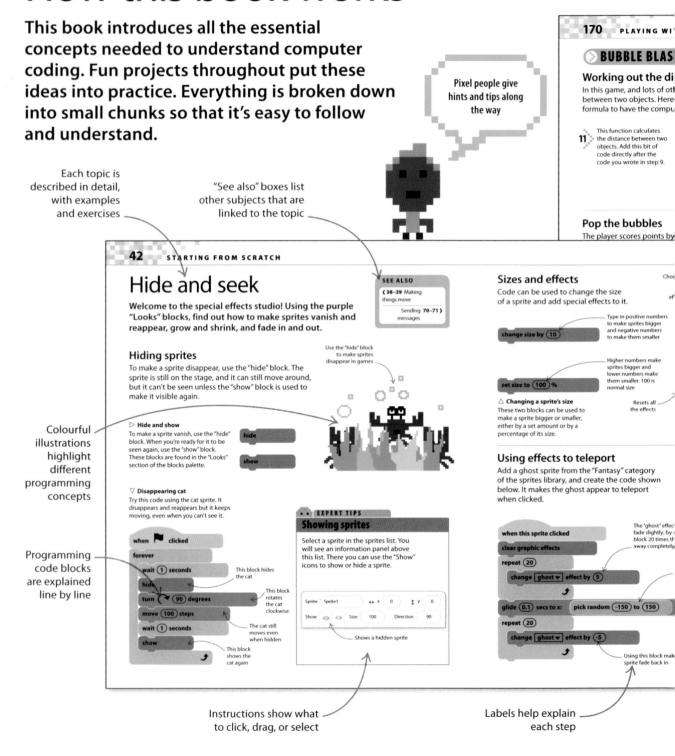

Each topic is described in detail, with examples and exercises

"See also" boxes list other subjects that are linked to the topic

Colourful illustrations highlight different programming concepts

Programming code blocks are explained line by line

170 PLAYING WI

> **BUBBLE BLAS**

Working out the di
In this game, and lots of oth between two objects. Here formula to have the compu

11 This function calculates the distance between two objects. Add this bit of code directly after the code you wrote in step 9.

Pop the bubbles
The player scores points by

---

42 STARTING FROM SCRATCH

Hide and seek

Welcome to the special effects studio! Using the purple "Looks" blocks, find out how to make sprites vanish and reappear, grow and shrink, and fade in and out.

SEE ALSO
‹ **38–39** Making things move
Sending **70–71** › messages

Hiding sprites
To make a sprite disappear, use the "hide" block. The sprite is still on the stage, and it can still move around, but it can't be seen unless the "show" block is used to make it visible again.

Use the "hide" block to make sprites disappear in games

▷ **Hide and show**
To make a sprite vanish, use the "hide" block. When you're ready for it to be seen again, use the "show" block. These blocks are found in the "Looks" section of the blocks palette.

`hide`

`show`

▽ **Disappearing cat**
Try this code using the cat sprite. It disappears and reappears but it keeps moving, even when you can't see it.

```
when [flag] clicked
forever
  wait (1) seconds
  hide
  turn ↻ (90) degrees
  move (100) steps
  wait (1) seconds
  show
```

This block hides the cat

This block rotates the cat clockwise

The cat still moves even when hidden

This block shows the cat again

• • **EXPERT TIPS**
Showing sprites

Select a sprite in the sprites list. You will see an information panel above this list. There you can use the "Show" icons to show or hide a sprite.

| Sprite | Sprite1 | ↔ x | 0 | ↕ y | 0 |
| Show | 👁 👁 | Size | 100 | Direction | 90 |

Shows a hidden sprite

Sizes and effects
Code can be used to change the size of a sprite and add special effects to it.

Choo ef

`change size by (10)`

Type in positive numbers to make sprites bigger and negative numbers to make them smaller

`set size to (100) %`

Higher numbers make sprites bigger and lower numbers make them smaller. 100 is normal size

△ **Changing a sprite's size**
These two blocks can be used to make a sprite bigger or smaller, either by a set amount or by a percentage of its size.

Resets all the effects

Using effects to teleport
Add a ghost sprite from the "Fantasy" category of the sprites library, and create the code shown below. It makes the ghost appear to teleport when clicked.

```
when this sprite clicked
clear graphic effects
repeat (20)
  change [ghost ▾] effect by (5)
glide (0.1) secs to x: pick random (-150) to (150)
repeat (20)
  change [ghost ▾] effect by (-5)
```

The "ghost" effec fade slightly; by block 20 times th away completely

Using this block mak sprite fade back in

Instructions show what to click, drag, or select

Labels help explain each step

Seven projects build up coding skills. Project pages are highlighted with a blue band

Simple step-by-step instructions guide you through each project

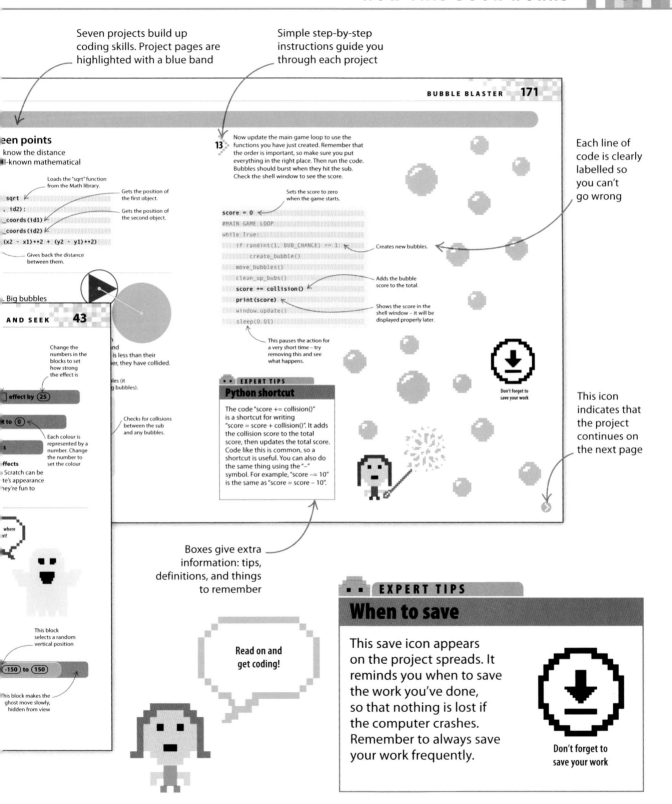

Each line of code is clearly labelled so you can't go wrong

een points

know the distance
ll-known mathematical

Loads the "sqrt" function from the Math library.

sqrt

Gets the position of the first object.

, id2) :

_coords(id1)

Gets the position of the second object.

_coords(id2)

(x2 - x1)**2 + (y2 - y1)**2)

Gives back the distance between them.

. Big bubbles

AND SEEK   43

Change the numbers in the blocks to set how strong the effect is

] effect by (25)

t to (0)

Each colour is represented by a number. Change the number to set the colour

ffects
Scratch can
e's appearance
hey're fun to

where
xt!

This block selects a random vertical position

(-150) to (150)

This block makes the ghost move slowly, hidden from view

13 Now update the main game loop to use the functions you have just created. Remember that the order is important, so make sure you put everything in the right place. Then run the code. Bubbles should burst when they hit the sub. Check the shell window to see the score.

Sets the score to zero when the game starts.

```
score = 0
#MAIN GAME LOOP
while True:
    if randint(1, BUB_CHANCE) == 1:
        create_bubble()
    move_bubbles()
    clean_up_bubs()
    score += collision()
    print(score)
    window.update()
    sleep(0.01)
```

Creates new bubbles.

Adds the bubble score to the total.

Shows the score in the shell window – it will be displayed properly later.

This pauses the action for a very short time – try removing this and see what happens.

is less than their
er, they have collided.

les (it
g bubbles).

Checks for collisions between the sub and any bubbles.

· · EXPERT TIPS
Python shortcut

The code "score += collision()" is a shortcut for writing "score = score + collision()". It adds the collision score to the total score, then updates the total score. Code like this is common, so a shortcut is useful. You can also do the same thing using the "–" symbol. For example, "score –= 10" is the same as "score = score – 10".

Don't forget to save your work

This icon indicates that the project continues on the next page

Boxes give extra information: tips, definitions, and things to remember

Read on and get coding!

· · EXPERT TIPS
When to save

This save icon appears on the project spreads. It reminds you when to save the work you've done, so that nothing is lost if the computer crashes. Remember to always save your work frequently.

Don't forget to save your work

What is coding?

What is a computer program?

A computer program is a set of instructions that a computer follows to complete a task. "Coding", or "programming", means writing the step-by-step instructions that tell the computer what to do.

SEE ALSO

Thinking like **16–17 〉**
a computer

Becoming **18–19 〉**
a coder

Computer programs are everywhere

We are surrounded by computer programs. Many of the devices and gadgets we use each day are controlled by them. These machines all follow step-by-step instructions written by a computer programmer.

◁ **Mobile phones**
Programs allow you to make a phone call or send text messages. When you search for a contact, a program finds the correct phone number.

△ **Computer software**
Everything a computer does, from browsing the internet to writing documents or playing music, works because of code written by a computer programmer.

◁ **Games**
Consoles are just another type of computer, and all the games that run on them are programs. All the graphics, sounds, and controls are written in computer code.

▷ **Cars**
In some cars, computer programs monitor the speed, temperature, and amount of fuel in the tank. Computer programs can even help control the brakes to keep people safe.

△ **Washing machines**
Washing machines are programmed to follow different cycles. Computer code controls how hot the water is and how long the wash takes.

How computer programs work

Computers might seem very smart, but they are actually just boxes that follow instructions very quickly and accurately. As intelligent humans, we can get them to carry out different tasks by writing programs, or lists of instructions.

1 Computers can't think
A computer won't do anything by itself. It's up to the computer programmer to give it instructions.

Without instructions a computer is clueless

This is a computer program counting down to launch

2 Write a program
You can tell a computer what to do by writing a set of very detailed instructions called a program. Each instruction has to be small enough that the computer can understand it. If the instructions are incorrect, the computer won't behave the way you want it to.

```
for count in range(10, 0, -1):
    print("Counting down", count)
```

3 Programming languages
Computers can only follow instructions in a language they understand. It's up to the programmer to choose which language is best for the task.

```
for count in range(10, 0, -1):
    print("Counting down", count)
```

All programs are finally converted into "binary code", a basic computer language that uses only ones and zeroes

```
0010 0011 1000 1100
1000 0110 0100 1001
0100 1001 0001 0101
```

BLAST OFF!

·:·· LINGO

Hardware and software

"Hardware" means the physical parts of the computer that you can see or touch (all the wires, the circuits, the keyboard, the display screen, and so on). "Software" means the programs that run on the computer and control how it works. Software and hardware work together to make computers do useful things.

Think like a computer

A programmer must learn to think like a computer. All tasks must be broken down into small chunks so they are easy to follow, and impossible to get wrong.

SEE ALSO

❮ **14–15** What is a computer program?

Becoming **18–19** ❯ a coder

Thinking like a robot

Imagine a café where the waiter is a robot. The robot has a simple computer brain, and needs to be told how to get from the café kitchen to serve food to diners seated at tables. First the process has to be broken down into simple tasks the computer can understand.

LINGO

Algorithm

An algorithm is a set of simple instructions for performing a task. A program is an algorithm that has been translated into a language that computers can understand.

1 **Waiter robot program 1**
Using this program the robot grabs the food from the plate, crashes straight through the kitchen wall into the dining area, and puts the food on the floor. This algorithm wasn't detailed enough.

1. Pick up food

2. Move from kitchen to diner's table

3. Put food down

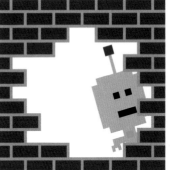

◁ **Disaster!**
The instructions weren't clear: we forgot to tell the robot to use the door. It might seem obvious to humans but computers can't think for themselves.

2 **Waiter robot program 2**
This time we've told the robot waiter to use the kitchen door. It makes it through the door, but then hits the café cat, trips, and smashes the plate on the floor.

1. Pick up a plate with food on it

2. Move from kitchen to diner's table by:

> Move to door between kitchen and dining area

> Move from door to the table

3. Put plate down on the table in front of the diner

△ **Still not perfect**
The robot doesn't know how to deal with obstacles like the cat. The program needs to give the robot even more detailed instructions so it can move around safely.

3 **Waiter robot program 3**
In this version of the program, the robot successfully delivers the food to the diner avoiding any obstacles. But after putting the plate down, the robot remains standing at the table while food piles up in the kitchen.

1. Pick up a plate with food on it holding it level at all times

2. Move from kitchen to diner's table by:

Move to door between kitchen and dining area

checking for obstacles and steering around them

Move from door to the table

checking for obstacles and steering around them

3. Put plate down on the table in front of the diner

△ **Success at last?**
Finally the robot can deliver the food safely. But we forgot to give it instructions to go back to the kitchen and get the next plate.

Real-world example

The waiter robot might be imaginary, but algorithms like this are in action all around us. For example, a computer-controlled lift faces the same sort of problems. Should it go up or down? Which floor should it go to next?

1. Wait until doors are closed

2. Wait for button to be pressed

If button pressed is higher than current floor:

Move lift upwards

If button pressed is lower than current floor:

Move lift downwards

3. Wait until current floor equals button pressed

4. Open doors

◁ **Lift program**
For the lift to work correctly and safely, every step has to be precise, clear, and cover every possibility. The programmers have to make sure they create a suitable algorithm.

Becoming a coder

Coders are the people who write the programs behind everything we see and do on a computer. You can create your own programs by learning a programming language.

SEE ALSO

What is **22–23 〉**
Scratch?

What is **86–87 〉**
Python?

Programming languages

There are a huge range of programming languages to choose from. Each one can be used for different tasks. Here are some of the most popular languages and what they are often used for:

C	A powerful language for building computer operating systems.

MATLAB	Ideal for programs that need to carry out lots of calculations.

Ada	Used to control spacecraft, satellites, and aeroplanes.

Ruby	Automatically turns lots of information into web pages.

Java	Works on computers, mobile phones, and tablets.

JavaScript	A language used to build interactive websites.

Scratch	A visual language that's ideal for learning programming. This is the first language covered in this book.

Python	A text-based language that can be used to build all kinds of things. It's the second language covered in this book.

What is Scratch?

Scratch is a great way to start coding. Programs are created by connecting together blocks of code, instead of typing it out. Scratch is quick and easy to use, and also teaches you the key ideas you need to use other programming languages.

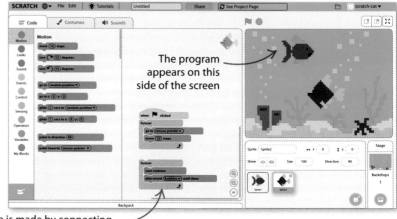

The program appears on this side of the screen

Code is made by connecting coloured blocks together

What is Python?

People around the world use Python to build games, tools, and websites. It's a great language to master as it can help you build all kinds of different programs. Python looks like a mixture of recognizable words and characters, so it can be easily read and understood by humans.

```
                              ghostgame
IDLE    File    Edit    Format    Run    Window    Help
# Ghost Game
from random import randint
print("Ghost Game")
feeling_brave = True
score = 0
while feeling_brave:
    ghost_door = randint(1, 3)
    print("Three doors ahead...")
```

A program written in Python

Getting started

It's time to start programming. All you need is a computer with an internet connection. This book starts with Scratch – the perfect language to help you on your way to becoming a coding expert. Get ready to jump into the exciting world of computer coding.

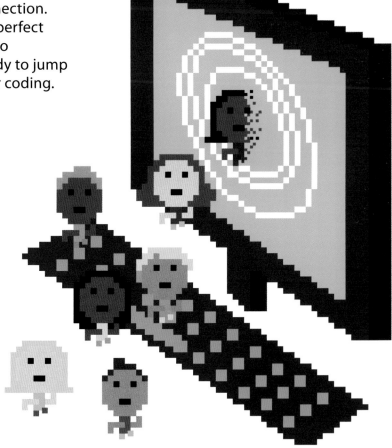

■ ■ EXPERT TIPS

Enjoy experimenting

As a programmer you should experiment with the code and programs you make. One of the best ways to learn programming is to play about and see what happens when you change different parts of the code. By tinkering and fiddling, you'll discover new ways of doing things. You'll learn much more about computer programming and have even more fun.

Starting from Scratch

What is Scratch?

Scratch is a visual programming language that makes coding simple. It can be used to make all sorts of fun and interesting programs.

SEE ALSO

Installing and **24–25 ⟩**
launching Scratch

Scratch **26–27 ⟩**
interface

Coloured blocks **30–31 ⟩**
and code

Understanding Scratch

Scratch is perfect for making games and animations. It has large collections (or "libraries") of cool graphics and sounds that you can play around with.

Blocks lock together like jigsaw pieces

1 Start programming
Scratch is a programming language. There's not much typing, and it's easy to get started.

Write your first program in Scratch!

2 Put together programming blocks
Scratch uses coloured blocks of code. Blocks are selected and joined together to make code, which is a set of instructions.

3 Make sprites move and speak
Objects such as people, vehicles, and animals can be added to a program. These objects are called sprites. Code blocks make them move and speak.

Sprites like me can be programmed to talk in speech bubbles.

Sprites can be programmed to walk, run, and dance

LINGO

Why is it called Scratch?

"Scratching" is a way of mixing different sounds to make new music. The Scratch programming language enables you to mix pictures, sounds, and code blocks to make new computer programs.

A typical Scratch program

Here is an example of a Scratch program. All of the action takes place in an area on the screen called the "stage". Background images and sprites can be added to the stage, and you can write code (sometimes called scripts) to make things happen.

▽ **Running a program**
Starting a program is called "running" it. To run a program in Scratch, click the green flag above the stage.

The green flag runs a program

The red button stops a program

Background image

Adding code makes the shark sprite move

Several sprites can be on the stage at once

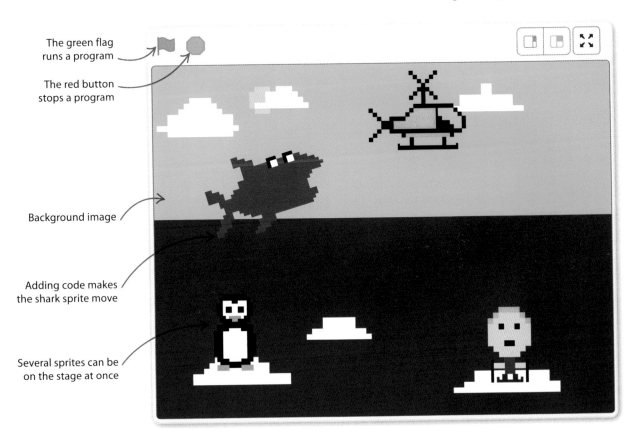

▷ **Code blocks make sprites move**
Scratch contains blocks that can be used to make code. This code makes the shark bounce around the screen. The "next costume" block makes it open and close its mouth with each movement.

The "forever" block keeps the sprite moving endlessly

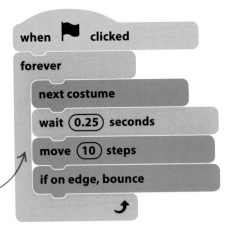

```
when [flag] clicked
forever
    next costume
    wait (0.25) seconds
    move (10) steps
    if on edge, bounce
```

Installing and launching Scratch

To start programming in Scratch, you need to have the Scratch software. It can be installed on a computer, or it can be used online.

Create a Scratch account

A Scratch account can be used to share the programs you make on the Scratch website. It's also used to save work online. Visit the Scratch website at: **http://scratch.mit.edu/** and click "Join Scratch" to create your account.

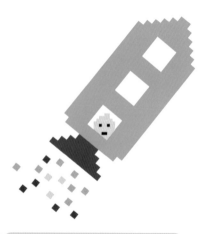

▷ **Getting started**
The way Scratch is set up depends on whether it's used over the internet (online) or from downloaded software (offline).

1 Set-up	**2 Launching Scratch**
Online — Visit **http://scratch.mit.edu** and click "Join Scratch". Fill in the form to create a username and password. Make sure you get permission from your parent or carer to join the website.	Once you've joined the Scratch website, click "Sign in", and enter your username and password. Click "Create" at the top of the screen to begin a new program.
Offline — Download the software version of Scratch at: **http://scratch.mit.edu/ download**. Run the installation program and a Scratch icon will appear on your desktop.	Double-click the icon on the desktop and Scratch will start, ready to begin programming.

Mouse control

The "click" instruction means press the left mouse button if there is more than one. "Right-click" means use the right mouse button. If a mouse only has one button, hold the "CTRL" key on the keyboard and press the mouse button to perform a right-click.

Different versions of Scratch

This book uses Scratch 3.0, the latest version of Scratch. Use this version if possible. An older version will differ slightly.

△ **Scratch 2.0**
The older version of Scratch has the stage on the left of the screen.

△ **Scratch 3.0**
The latest version of Scratch has some new commands and the stage is on the right of the screen.

3 Saving work

When you're logged in, Scratch automatically saves work for you. To find your work, click your username at the top right of the screen and click "My Stuff".

Click the "File" menu at the top of the screen and choose "Save to your computer". Ask the person who owns the computer where you should save your work.

4 Operating systems

The web version of Scratch 3.0 works well on Windows, Ubuntu, and Mac computers. This version will also work on tablets.

The offline version of Scratch works well on Windows and Mac computers. It doesn't work well on computers that use Ubuntu. If a computer uses Ubuntu, try the online version instead.

Ready? Let's go!

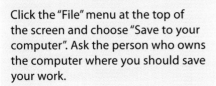

Scratch interface

This is Scratch's screen layout, or "interface". The stage is on the right and programs are created in the middle.

Change language

Menu options

Costumes tab

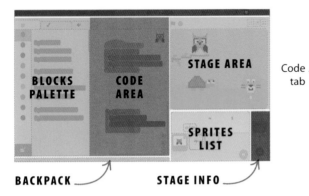

Code tab

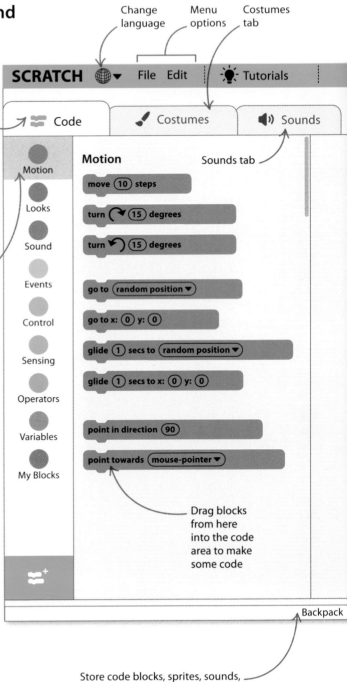

Sounds tab

Select different types of blocks

Drag blocks from here into the code area to make some code

△ **Scratch map**
The stage is where programs run. Sprites are managed in the sprite list and code blocks can be found in the blocks palette. Build code in the code area.

Store code blocks, sprites, sounds, and costumes in the backpack

• • EXPERT TIPS

Menu options

This is what the menu options at the top of the screen do.

File	**Save work** or start a new project.
Edit	**Undo any mistakes** or speed up the time between code blocks.
Tutorials	**If you get stuck,** find help here.

▽ **Experiment**
Click the buttons and tabs to explore and experiment with the Scratch interface. The projects that follow explain how to use them.

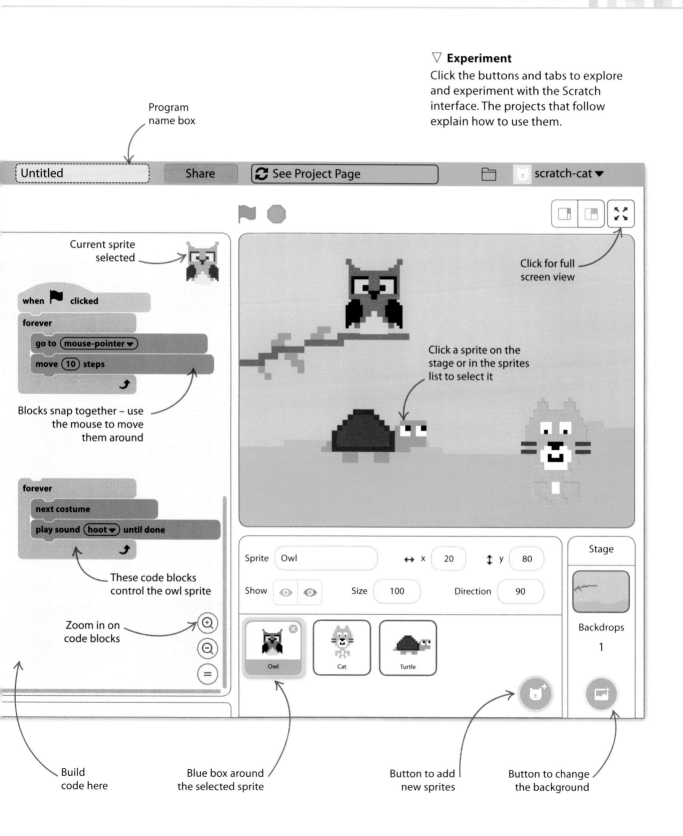

Program name box

Untitled   Share   ⟳ See Project Page   scratch-cat ▼

Current sprite selected

Click for full screen view

when ⚑ clicked
forever
  go to (mouse-pointer ▼)
  move (10) steps

Click a sprite on the stage or in the sprites list to select it

Blocks snap together – use the mouse to move them around

forever
  next costume
  play sound (hoot ▼) until done

These code blocks control the owl sprite

Zoom in on code blocks

Sprite Owl   ↔ x 20   ↕ y 80
Show 👁 ◑   Size 100   Direction 90

Owl   Cat   Turtle

Stage

Backdrops
1

Build code here

Blue box around the selected sprite

Button to add new sprites

Button to change the background

Sprites

Sprites are the basic components of Scratch. Every Scratch program is made up of sprites and the code that controls them. The "Escape the dragon!" program on pages 32–37 uses the cat, dragon, and donut sprites.

SEE ALSO

❮ **26–27** Scratch interface

Costumes **40–41** ❯

Hide and seek **42–43** ❯

What can sprites do?

Sprites are the images on the stage. Code blocks are programmed to make them do things. Sprites can be instructed to react to other sprites and the user of the program. Here are a few things sprites can do:

We can make lots of different sounds.

Move around the stage
Change their appearance
Play sounds and music

React when they touch things
Be controlled by the user
Talk in speech bubbles

Sprites in the Scratch interface

Each project can have several sprites, and each one can have its own code. It's important to add code to the correct sprite, and to know how to switch between them.

The code being shown belongs to the sprite shown here

Select different types of blocks by clicking on these icons

Select different sprites by clicking on them

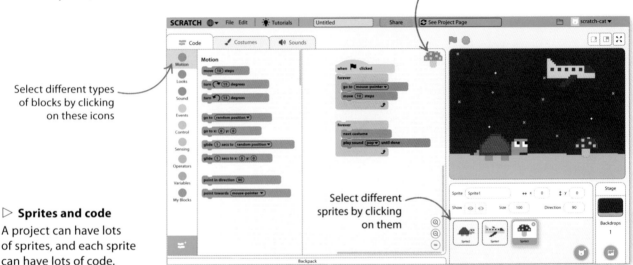

▷ **Sprites and code**

A project can have lots of sprites, and each sprite can have lots of code.

Creating and editing sprites

Games are more exciting when there are more sprites to hit, dodge, or chase each other around the stage. It's simple to create, copy, and delete sprites.

▽ Create a sprite

Select "Choose a Sprite" in the sprites list and use the buttons in the pop-up menu to add or create a sprite for your program.

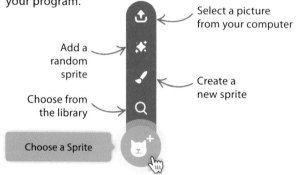

Select a picture from your computer

Add a random sprite

Create a new sprite

Choose from the library

Choose a Sprite

▽ Copy or delete a sprite

To copy a sprite and its code, right-click on it in the sprites list and choose "duplicate".

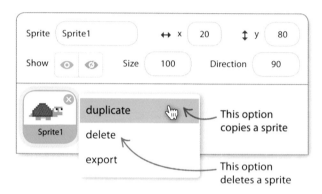

This option copies a sprite

This option deletes a sprite

Naming a sprite

When you start a new program in Scratch the cat sprite is called "Sprite1". It's easier to write programs if you give your sprites more meaningful names. It also makes it easier to understand and manage their code.

1 Select the sprite

Select a sprite in the sprites list by clicking on it.

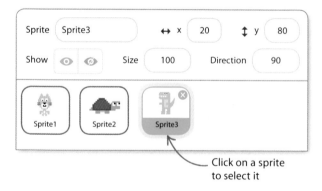

Click on a sprite to select it

2 Change the name

In the information panel, click on the text box and use the keyboard to change the name of the sprite. The sprite has now been renamed.

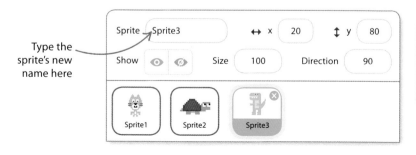

Type the sprite's new name here

The sprite's new name appears in the sprites list

Coloured blocks and code

SEE ALSO

❮ **26–27** Scratch interface

Escape the **32–37** ❯ dragon!

Blocks are colour-coded depending on what they do. Putting them together builds code that runs in the order in which they are placed.

Coloured blocks

There are nine different types of blocks in Scratch. Switch between them using the buttons in the blocks palette. Click on a colour to see all the blocks in that section.

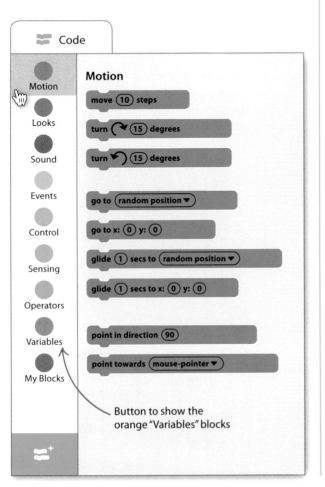

Button to show the orange "Variables" blocks

Functions of blocks

Different types of blocks do different things in programs. Some of them make sprites move, some manage sounds, and some decide when things happen.

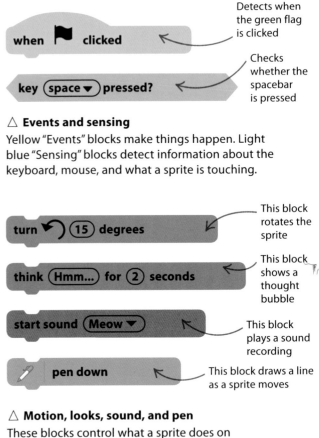

Detects when the green flag is clicked

Checks whether the spacebar is pressed

△ **Events and sensing**
Yellow "Events" blocks make things happen. Light blue "Sensing" blocks detect information about the keyboard, mouse, and what a sprite is touching.

This block rotates the sprite

This block shows a thought bubble

This block plays a sound recording

This block draws a line as a sprite moves

△ **Motion, looks, sound, and pen**
These blocks control what a sprite does on screen – this is called the output of a program. Pick a sprite and try each block to see what it does.

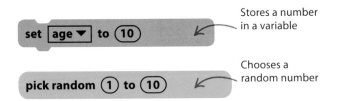

Stores a number
in a variable

Chooses a
random number

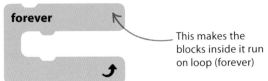

This makes the
blocks inside it run
on loop (forever)

△ **Variables and operators**
Orange "Variables" blocks and green
"Operators" blocks store numbers
and words and do things with them.

△ **Control**
The "Control" blocks make decisions about
when blocks run. They can be programmed
to repeat instructions.

Flow of code

When a program runs, Scratch carries out the
instructions on the blocks. It starts at the top
of the code and works its way down.

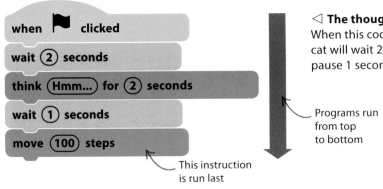

This instruction
is run last

Programs run
from top
to bottom

◁ **The thoughtful cat**
When this code is used with the cat sprite, the
cat will wait 2 seconds, think for a moment,
pause 1 second, and then move.

Running the code

When the code is running, it glows. Use the green
flag button on the stage to run the code or click
a line or block of code to make it run.

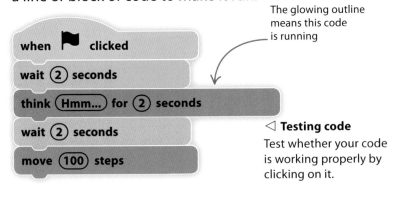

The glowing outline
means this code
is running

◁ **Testing code**
Test whether your code
is working properly by
clicking on it.

⣿ REMEMBER

Stopping the code

To stop any code blocks in a
program that are running, click the
red stop button above the stage.
It's shaped like an octagon. You'll
find it beside the green flag button
used to start your program.

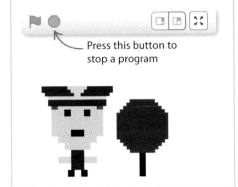

Press this button to
stop a program

> PROJECT 1

Escape the dragon!

This project introduces some basic Scratch coding. It shows how to make a game to help the cat sprite dodge a fire-breathing dragon.

SEE ALSO

❮ **24-25** Installing and launching Scratch

❮ **26–27** Scratch interface

Make the cat move

This stage explains how to make the cat sprite move around and chase the mouse-pointer. Follow the instructions carefully, otherwise the game might not work.

1 Open Scratch. Click "File" on the menu and select "New" to start a new project. The cat sprite appears.

> Every new project in Scratch includes me, the cat sprite.

2 Click the orange "Control" button in the blocks palette. Then click the "forever" block, keep the mouse button pressed down, and drag the block into the code area on the right. Release the button to drop the block.

Blocks palette

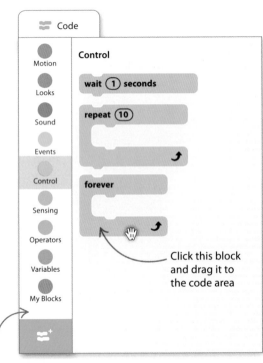

Click this block and drag it to the code area

3 Click the blue "Motion" button in the blocks palette. The blue "Motion" commands will appear. Drag the "point towards" block into the code area and drop it inside the "forever" block.

Drop this block inside the "forever" block

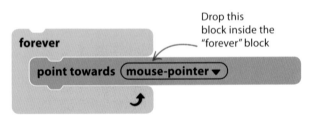

4 Click the "Events" button in the blocks palette. Drag the "when green flag clicked" block into the code area. Join it to the top of your code.

This block snaps to the top of the code

The menu shows "mouse-pointer" has been chosen

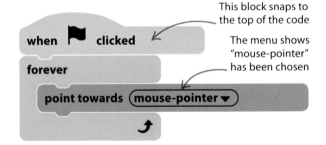

5 Try running the program by clicking the green flag at the top of the stage. As you move the mouse around the stage, the cat turns to face the mouse-pointer.

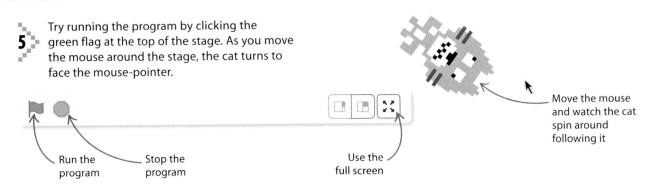

Run the program

Stop the program

Use the full screen

Move the mouse and watch the cat spin around following it

6 Click the "Motion" button again, and drag the "move 10 steps" block into the code area. Drop it inside the "forever" block. Click the green flag button so the cat chases the mouse-pointer!

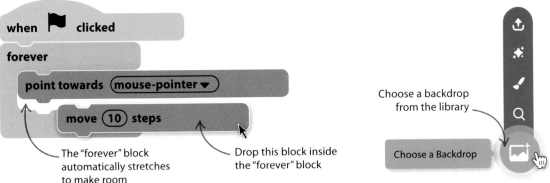

The "forever" block automatically stretches to make room

Drop this block inside the "forever" block

7 The picture behind the sprites is called a backdrop. To the right of the sprites list is a button to add a backdrop from the library. Click it, select the "Space" theme from the list, and then click the "Stars" image to select this backdrop.

Choose a backdrop from the library

Choose a Backdrop

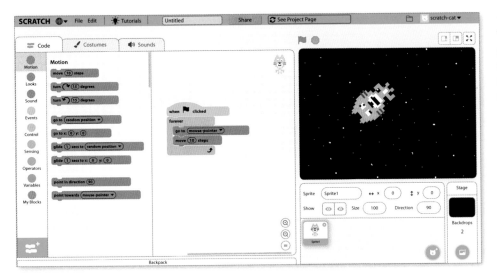

◁ **Cat in space**
The Scratch interface now looks like this. Run the program and the cat chases the mouse-pointer through space.

Scratch automatically saves work if you're online. To save work while offline – click "File" and select "Save to your computer".

⊙ ESCAPE THE DRAGON!

Add a fire-breathing dragon

Now that the cat can chase the mouse, make a dragon to chase the cat. Don't let the dragon catch the cat, or it will get scorched.

8 Below the sprites list is a button to add a sprite from the library. Click it, choose the "Fantasy" category from the menu on top, and select "Dragon".

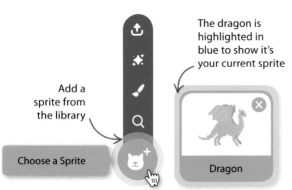

Add a sprite from the library

Choose a Sprite

The dragon is highlighted in blue to show it's your current sprite

Dragon

9 Add this code to the dragon sprite. Click the colour-coded buttons in the blocks palette to select the blocks below, then drag them into the code area. The dragon will now chase the cat.

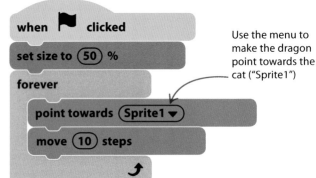

```
when [flag] clicked
set size to (50) %
forever
    point towards (Sprite1 ▼)
    move (10) steps
```

Use the menu to make the dragon point towards the cat ("Sprite1")

10 Click the blue "Motion" button and drag the "go to x:0 y:0" block into the code. Click the number boxes in the block and change them to -200 and -150. Click the purple "Looks" button and add the "switch costume to" block to your code.

```
when [flag] clicked
set size to (50) %
go to x: (-200) y: (-150)
switch costume to (dragon-a ▼)
forever
    point towards (Sprite1 ▼)
    move (4) steps
```

Place this block here to make the dragon start in the corner

Use the menu to choose "dragon-a". The dragon will start in this costume

Change 10 to 4 to make the dragon move slower than the cat

11 With the dragon sprite highlighted, add this second bit of code to the code area. The "wait until" block is found in the "Control" section, and the "touching" block is in the "Sensing" section. The dragon now breathes fire when it touches the cat.

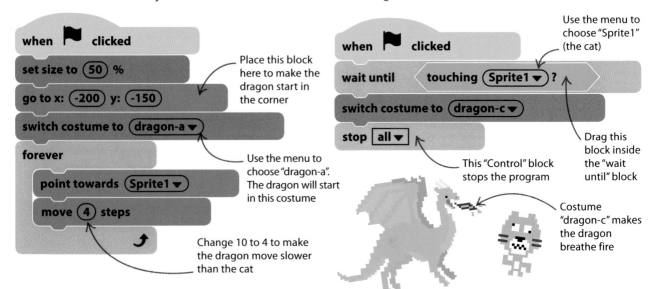

```
when [flag] clicked
wait until ⟨ touching (Sprite1 ▼) ? ⟩
switch costume to (dragon-c ▼)
stop (all ▼)
```

Use the menu to choose "Sprite1" (the cat)

Drag this block inside the "wait until" block

This "Control" block stops the program

Costume "dragon-c" makes the dragon breathe fire

12 In coding, a "variable" is used to store information. This step uses a variable to create a timer to measure how long a player survives before getting toasted. Click the "Variables" button and then click "Make a Variable".

13 Type in the variable name "Time" and make sure the "For all sprites" button is selected underneath, then click "OK". This means that the cat, dragon, and any other sprites can use the variable.

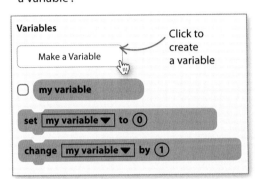

Click to create a variable

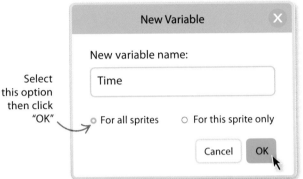

Select this option then click "OK"

14 The variable name and the number in it appear on the stage in a small box. Right-click it and choose "large readout". This shows just the number in the box.

The number in your "Time" variable

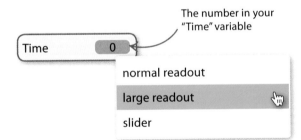

EXPERT TIPS

Make the game harder

Try changing the speed or size of your sprites.

Make the dragon faster:

move (5) steps

Make the dragon larger or smaller:

Change the value inside the number box to make a sprite larger or smaller. See how big you can make each sprite.

Size 100

15 Making a variable adds new blocks to the "Variables" section of the blocks palette. Drag the "set my variable to 0" and "change my variable by 1" blocks from the "Variables" section to the code area to make this new code. Click the drop-down menu in both blocks and choose "Time". You can give this code to any sprite.

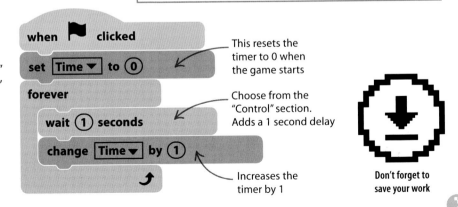

This resets the timer to 0 when the game starts

Choose from the "Control" section. Adds a 1 second delay

Increases the timer by 1

Don't forget to save your work

⊙ ESCAPE THE DRAGON!

Add a delicious donut

Scratch comes with lots of sprites in its library. Make the game trickier by adding a donut sprite to the program for the cat to chase.

16 Click the button below the sprites list to add a new sprite from the library. Search for "Donut" and select it.

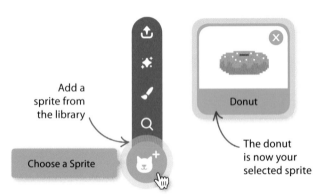

Add a sprite from the library

Choose a Sprite

The donut is now your selected sprite

Donut

17 Add this code to the donut. The "mouse down?" block can be found in the "Sensing" section, and the "go to mouse-pointer" block in the "Motion" section. This code makes the donut follow the mouse-pointer when the mouse button is clicked.

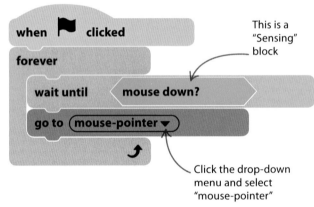

when ⚑ clicked

forever

wait until   mouse down?

This is a "Sensing" block

go to mouse-pointer ▼

Click the drop-down menu and select "mouse-pointer"

18 Select the cat in the sprites list so its code appears. Click the arrow in the "point towards mouse-pointer" block and choose "Donut". Do this so that the cat follows the donut instead of the mouse-pointer.

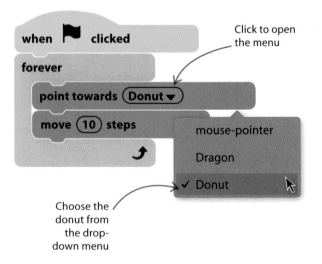

when ⚑ clicked

forever

point towards Donut ▼

move 10 steps

Click to open the menu

mouse-pointer

Dragon

✓ Donut

Choose the donut from the drop-down menu

19 Click the green flag button to run the program. Press the mouse button and the donut moves to the mouse-pointer. The cat follows the donut, and the dragon chases the cat.

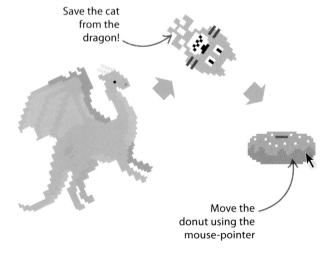

Save the cat from the dragon!

Move the donut using the mouse-pointer

20 Now add some music. Click the "Sounds" tab above the blocks palette. Each sprite has its own sounds, and they are managed here. Click the button on the bottom-left to add a sound from the library.

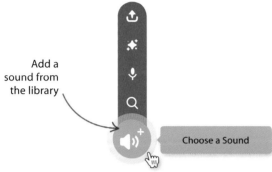

Add a sound from the library

Choose a Sound

21 Search for the "Drip Drop" sound and select it. The sound is added to the cat sprite, and appears in the "Sounds" area.

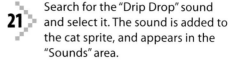

Delete sounds here

Drip Drop 2.86

This is how long the sound lasts

22 Click the "Code" tab to go back to the code area. Add this code to the cat sprite, so it plays the music all the time. Run the program and have fun!

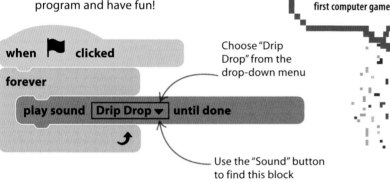

when 🏳 clicked

forever

play sound Drip Drop ▼ until done

Choose "Drip Drop" from the drop-down menu

Use the "Sound" button to find this block

Congratulations! You've written your first computer game

Don't forget to save your work

∷∷ REMEMBER

Achievements

This project has shown some of the things Scratch can do. Here's what you've achieved.

Created a program: By combining blocks of code, you've put together a game.

Added pictures: You've used both backdrops and sprites.

Made sprites move: You've made sprites chase each other.

Used a variable: You've created a timer for your game.

Used costumes: You've changed the dragon's appearance using different costumes.

Added music: You've added a sound, and made it play when your program runs.

Making things move

Computer games are all about firing, dodging, catching, and escaping. Characters might run, fly spaceships, or drive fast cars. To create great games in Scratch, you first need to learn how to make sprites move.

SEE ALSO

‹ **28–29** Sprites

Co-ordinates **56–57** ›

Motion blocks

The dark blue "Motion" blocks make sprites move. Start a new project by clicking the "File" menu and choosing "New". The new project begins with the cat in the middle of the stage, ready for action.

1 **First steps**

Drag the "move 10 steps" block from the "Motion" section of the blocks palette and drop it into the code area to its right. Drag an orange "forever" block from the "Control" section of the blocks palette and drop it around this block. Click the green flag on the stage to run the program. The cat moves until it hits the edge of the stage.

Add this block to tell Scratch when to start running the code

Click the white window on the block and type in a different number to change how far the cat moves

The "forever" block repeats anything inside it endlessly

2 **Bouncing**

Drag an "if on edge, bounce" block inside your "forever" block. Now the cat bounces when it hits the edge of the stage. The cat is upside down when it walks to the left.

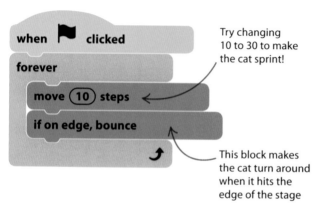

Try changing 10 to 30 to make the cat sprint!

This block makes the cat turn around when it hits the edge of the stage

3 **Rotating**

Drag the blue "set rotation style" block into the "forever" block and drop it below the "if on edge, bounce" block. Now the cat will face the right way up after bouncing off the edge of the stage.

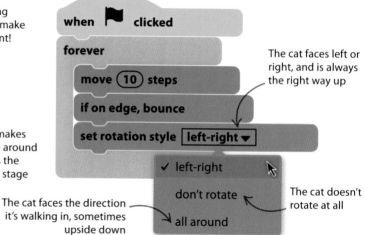

The cat faces left or right, and is always the right way up

The cat doesn't rotate at all

The cat faces the direction it's walking in, sometimes upside down

Which direction?

The cat is now marching left and right across the screen. It's possible to change the cat's direction, so it walks up and down, or even diagonally. The "Motion" blocks can be used to make a game of cat and mouse.

The direction -90° means "left"

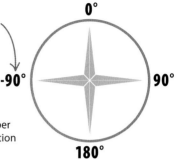

△ **Compass**
Directions are measured in degrees, from 0° at the top. You can use any number between -179° and +180°.

4 **Heading the right way**

Click the block to make the cat change direction

Type in a new number to change the direction of the cat

point in direction (-90)

Drag the "point in direction" block into the code area. Click on the number in the white window and type in a new direction.

Move the arrow on the drop-down compass to select a direction for the cat

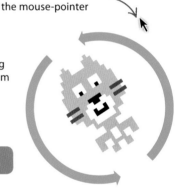

The cat will follow the mouse-pointer

5 **Cat and mouse**

Remove the "move 10 steps" and "if on edge, bounce" blocks from the code. Now drag a "point towards" block into the "forever" block. Open the menu and choose "mouse-pointer".

Click the green flag to start the program

when 🏴 clicked

forever

    point towards (mouse-pointer ▼)

As the mouse-pointer moves, the cat turns to face it

6 **Chase the mouse**

Can the cat catch the mouse? Drag a "move 10 steps" block into the "forever" loop. Now the cat walks towards the mouse-pointer.

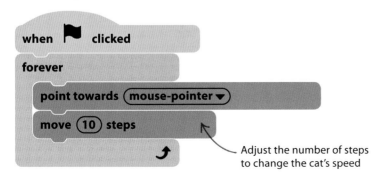

when 🏴 clicked

forever

    point towards (mouse-pointer ▼)

    move (10) steps

Adjust the number of steps to change the cat's speed

⣿ **REMEMBER**

Sprites

Sprites are objects in a Scratch program that you can move around (see pp.28–29). Every new project stars the cat sprite, but you can add cars, dinosaurs, dancers, and all sorts of other sprites from the library. You can even have a go at designing your own.

Costumes

To change what a sprite looks like, its expression, or its position, you need to change its "costume". Costumes are pictures of a sprite in different poses.

SEE ALSO

❮ **38–39** Making things move

Sending **70–71** ❯ messages

Changing costumes

Different costumes can make your sprite look like it's moving its arms and legs. When you switch between the cat's two costumes, it looks like it's walking. Start a new project and try this example.

One of the cat's costumes

1 **Different costumes**
Click the "Costumes" tab to see the cat's costumes. They show the cat with its legs and arms in two different positions.

2 **Make the cat walk**
Add this code to make the cat walk. When it moves, it slides across the screen without moving its legs, because its picture always stays the same.

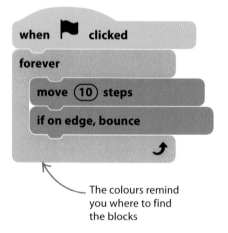

The colours remind you where to find the blocks

3 **Change the cat's costume**
Add the "next costume" block from the "Looks" section of the blocks palette, so the cat changes its costume with each step. This makes its legs and arms move.

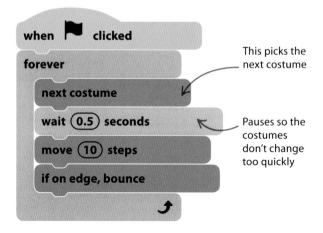

This picks the next costume

Pauses so the costumes don't change too quickly

Dancing ballerina

Now try making a ballerina dance. Add the ballerina sprite from the library. Select your cat in the sprites list and drag its code on to the ballerina in the sprites list. This copies the code to the ballerina.

Drop the code on to the ballerina in the sprites list

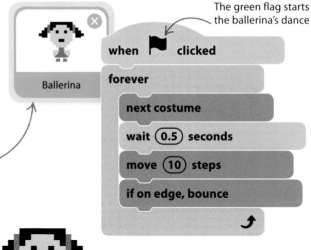

The green flag starts the ballerina's dance

when ⚑ clicked

forever
> next costume
> wait (0.5) seconds
> move (10) steps
> if on edge, bounce

△ **Ballerina's code**
The same code works for the ballerina and the cat. The ballerina has four costumes, and she uses them all as she dances on the stage.

Switching

You can choose to show a specific costume for your sprite using the "switch costume to" block. You can use this block to choose a particular position for your sprite.

switch costume to (ballerina-a ▼)

Switch costumes: Use the menu in the block to choose a costume.

switch backdrop to (backdrop1 ▼)

Switch backdrops: Change the picture on the stage with this block.

Adding speech bubbles

You can add speech bubbles to make your sprites talk when they change costumes. Use the "say Hello! for 2 seconds" block and change the text in it to make your sprite say something else.

The ballerina says "Up!"

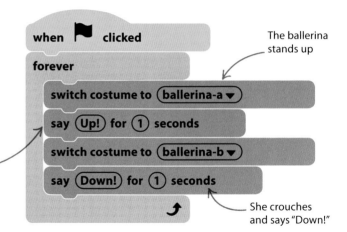

when ⚑ clicked

forever
> switch costume to (ballerina-a ▼)
> say (Up!) for (1) seconds
> switch costume to (ballerina-b ▼)
> say (Down!) for (1) seconds

The ballerina stands up

She crouches and says "Down!"

Hide and seek

Welcome to the special effects studio! Using the purple "Looks" blocks, find out how to make sprites vanish and reappear, grow and shrink, and fade in and out.

SEE ALSO

❰38–39 Making things move

Sending **70–71 ❱** messages

Hiding sprites

To make a sprite disappear, use the "hide" block. The sprite is still on the stage, and it can still move around, but it can't be seen unless the "show" block is used to make it visible again.

Use the "hide" block to make sprites disappear in games

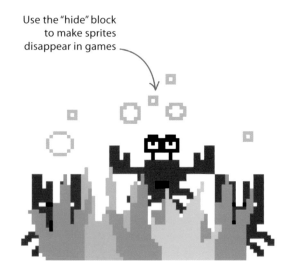

▷ **Hide and show**

To make a sprite vanish, use the "hide" block. When you're ready for it to be seen again, use the "show" block. These blocks are found in the "Looks" section of the blocks palette.

▽ **Disappearing cat**

Try this code using the cat sprite. It disappears and reappears but it keeps moving, even when you can't see it.

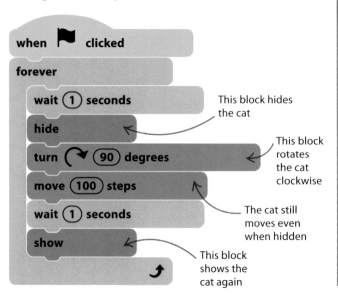

when ⚑ clicked

forever

wait ① seconds — This block hides the cat

hide

turn ↻ 90 degrees — This block rotates the cat clockwise

move 100 steps — The cat still moves even when hidden

wait ① seconds

show — This block shows the cat again

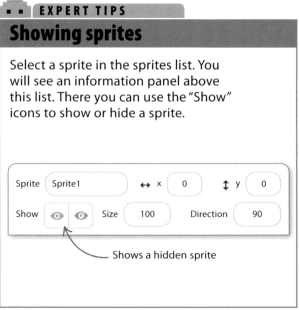

EXPERT TIPS

Showing sprites

Select a sprite in the sprites list. You will see an information panel above this list. There you can use the "Show" icons to show or hide a sprite.

| Sprite | Sprite1 | ↔ x | 0 | ↕ y | 0 |
| Show | 👁 🚫 | Size | 100 | Direction | 90 |

Shows a hidden sprite

Sizes and effects

Code can be used to change the size of a sprite and add special effects to it.

Choose the type of effect from the drop-down menu. The "pixelate" effect makes the sprite become blurred

Change the numbers in the blocks to set how strong the effect is

Type in positive numbers to make sprites bigger and negative numbers to make them smaller

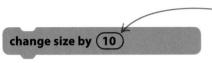
change size by (10)

change | pixelate ▼ | effect by (25)

set | color ▼ | effect to (0)

Higher numbers make sprites bigger and lower numbers make them smaller. 100 is normal size

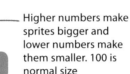
set size to (100) %

clear graphic effects

Each colour is represented by a number. Change the number to set the colour

△ **Changing a sprite's size**
These two blocks can be used to make a sprite bigger or smaller, either by a set amount or by a percentage of its size.

Resets all the effects

△ **Adding graphic effects**
The graphic effects in Scratch can be used to change a sprite's appearance or distort its shape. They're fun to experiment with.

Using effects to teleport

Add a ghost sprite from the "Fantasy" category of the sprites library, and create the code shown below. It makes the ghost appear to teleport when clicked.

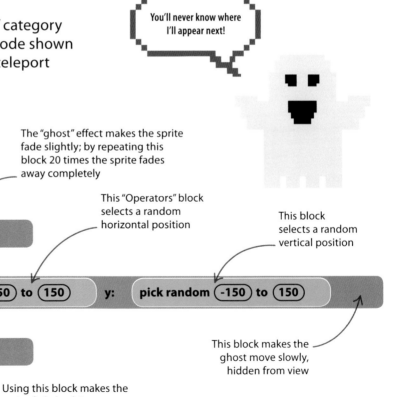

You'll never know where I'll appear next!

when this sprite clicked

clear graphic effects

repeat (20)
  change | ghost ▼ | effect by (5)

The "ghost" effect makes the sprite fade slightly; by repeating this block 20 times the sprite fades away completely

This "Operators" block selects a random horizontal position

This block selects a random vertical position

glide (0.1) secs to x: (pick random (-150) to (150)) y: (pick random (-150) to (150))

repeat (20)
  change | ghost ▼ | effect by (-5)

This block makes the ghost move slowly, hidden from view

Using this block makes the sprite fade back in

Events

The yellow "Events" blocks in Scratch start code when certain things happen. For example, when the user presses a key, clicks a sprite, or uses a webcam or microphone.

SEE ALSO

Sensing **66–67 〉**
and detecting

Sending **70–71 〉**
messages

Clicking

Code can be added to a sprite that makes it do something if the sprite is clicked while the program is running. Experiment with different blocks to see what a sprite can do when clicked.

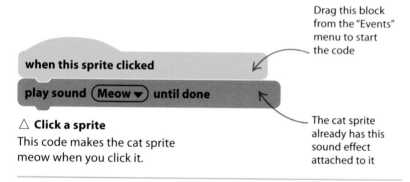

Drag this block from the "Events" menu to start the code

when this sprite clicked

play sound (Meow ▼) until done

The cat sprite already has this sound effect attached to it

△ **Click a sprite**
This code makes the cat sprite meow when you click it.

LINGO
What is an event?

An event is something that happens, such as a key being pressed or the green flag being clicked. The blocks that look for events go at the top of the code. The code waits until the event happens, and then it runs.

Key presses

Programs can be built to react when different keys on the keyboard are pressed. For another way of using the keyboard that's better for creating games, see pages 66–67.

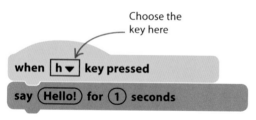

Choose the key here

when [h ▼] key pressed

say (Hello!) for (1) seconds

△ **Say hello**
Add this code to a sprite and when the H key is pressed, the sprite says "Hello!"

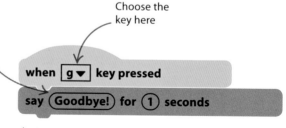

Choose the key here

Change the text here

when [g ▼] key pressed

say (Goodbye!) for (1) seconds

△ **Say goodbye**
This code uses the G key to make a sprite say "Goodbye!"

Sound events

If your computer has a microphone, sprites can detect how loud the sounds in a room are on a scale of 0 (very quiet) to 100 (very loud). Use the "when loudness > 10" block to make the code start when the sounds are loud enough.

1 Make the cat sensitive to noise
Start a new project, and add the "Room 2" backdrop image from the backdrop library. Drag the cat sprite on to the chair and add the code shown here.

Change the number to 40

```
when  loudness ▼  >  40

go to x: 145  y: 130

play sound  Meow ▼  until done

go to x: 145  y: 0
```

This makes the cat jump up

This makes the cat fall back down

2 Shout at the cat
Shout into the microphone – the cat will jump out of its seat with fright and meow. It will also respond to music and other sounds if they are loud enough.

Webcam motion detector

If you have a webcam, it can be used with Scratch too. Add this code to the cat, and when you wave at it through the webcam, it will meow back. To use the webcam blocks, click "Add Extension" at the bottom left then choose "Video Sensing".

Change the number to 40

```
when video motion >  40

play sound  Meow ▼  until done
```

△ **Detect motion**
Use the "when video motion > 10" block. The code will start when you're moving around enough.

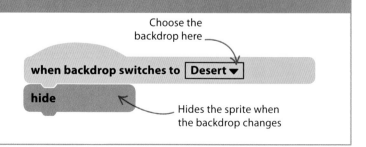

Simple loops

A loop is a part of a program that repeats itself. The loop blocks (from the "Control" section) tell Scratch which blocks to repeat, and how many times. They save us from adding the same blocks over and over again.

SEE ALSO

Complex **68–69 〉** loops

Loops **122–123 〉** in Python

Forever loop

Whatever you put inside the "forever" block repeats itself forever. There's no option to join anything at the bottom, because a "forever" loop never ends.

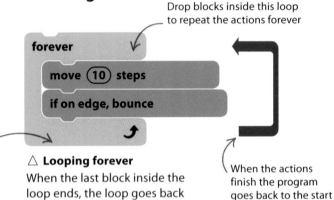

Drop blocks inside this loop to repeat the actions forever

No option to add more Scratch blocks

△ **Looping forever**
When the last block inside the loop ends, the loop goes back to the start again.

When the actions finish the program goes back to the start of the loop again

Repeat loop

To repeat an action a certain number of times, use a "repeat 10" block. Change the number in it to set how many times the loop will repeat itself. Add the "Dinosaur4" sprite to a new project and build it this code.

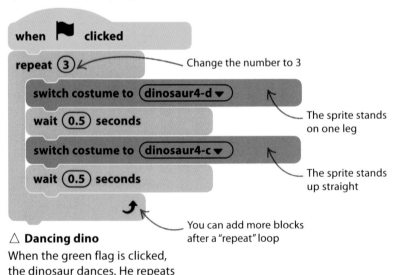

Change the number to 3

The sprite stands on one leg

The sprite stands up straight

You can add more blocks after a "repeat" loop

△ **Dancing dino**
When the green flag is clicked, the dinosaur dances. He repeats his dance moves three times.

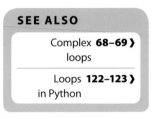 **REMEMBER**

Loop block shape

The loop blocks are shaped like jaws. Drop the blocks that you want to repeat into the jaws, so the loop wraps around them. As you add more blocks, the jaws stretch to make room for them.

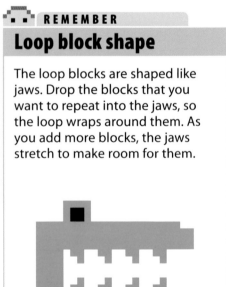

Nested loops

Loops can also be "nested", which means they can be put inside each other. In this code, the dinosaur finishes his dance by walking right and left and then thinking for a moment. When he's got his breath back, he dances again and stops only when you click the red stop button.

Try giving me some looping music!

Click on the green flag above the stage to start the code

▷ **Loops in loops**
This "forever" loop has several repeat loops inside it. Make sure the blocks are inside the right loops, otherwise the program won't work properly.

The "forever" block surrounds everything

The previous dance move (see opposite)

The dinosaur moves three steps to the right

The dinosaur moves three steps to the left

when ⚑ clicked

forever

repeat (3)

switch costume to (dinosaur4-d ▼)

wait (0.5) seconds

switch costume to (dinosaur4-c ▼)

wait (0.5) seconds

repeat (3)

move (20) steps

wait (0.5) seconds

wait (1) seconds

repeat (3)

move (-20) steps

wait (0.5) seconds

think (I love to dance!) for (2) seconds

This block creates a short pause

Type in what you want the dinosaur to think – it will appear in a thought bubble

Pens and turtles

Each sprite has a pen tool that can draw a line behind it wherever it goes. To create a picture, turn on the pen and then move the sprite across the stage, like moving a pen across paper. To use the pen blocks, click "Add Extension" at the bottom left then choose "Pen".

SEE ALSO

❮ **44–45** Events

❮ **46–47** Simple loops

Pen blocks

The dark green blocks are used to control the pen. Each sprite has its own pen that can be turned on by using the "pen down" block and turned off using the "pen up" block. The size and colour of the pen can also be changed.

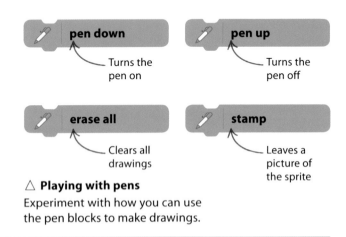

pen down — Turns the pen on

pen up — Turns the pen off

erase all — Clears all drawings

stamp — Leaves a picture of the sprite

△ **Playing with pens**
Experiment with how you can use the pen blocks to make drawings.

Draw a square

To draw a square, you simply put the pen down on the stage and then move the sprite in a square shape. Use a loop to draw the four sides and turn the corners.

The sprite will leave a line behind it

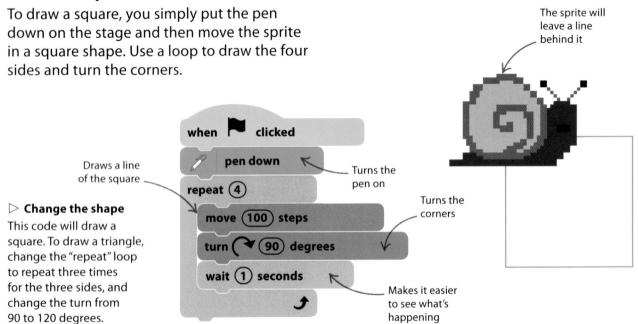

▷ **Change the shape**
This code will draw a square. To draw a triangle, change the "repeat" loop to repeat three times for the three sides, and change the turn from 90 to 120 degrees.

when ⚑ clicked

pen down — Turns the pen on

repeat 4

Draws a line of the square →

move 100 steps

turn ↻ 90 degrees — Turns the corners

wait 1 seconds — Makes it easier to see what's happening

Skywriting

In this program, you control a plane. As you fly it will leave
a smoke trail, so you can draw in the sky. Start a new project
and upload a plane sprite, then add this code.

You can only use
colours that appear on
the Scratch interface.
To select red, click in
the oval and then
move the button for
"Color" to the extreme
left or extreme right
to choose red

▷ **Flying high**
Use the left and right keys to turn the
plane. Switch on the smoke with the "a"
key and turn it off with the "z" key. Press
the spacebar to clear the sky.

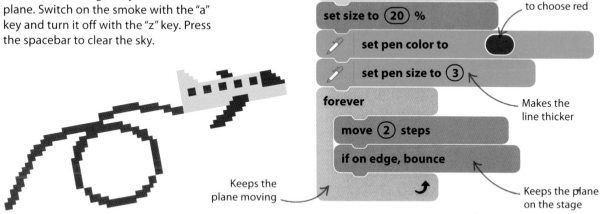

```
when 🏳 clicked
set size to (20) %
🖊 set pen color to ⬤
🖊 set pen size to (3)
forever
    move (2) steps
    if on edge, bounce
```

Makes the
line thicker

Keeps the
plane moving

Keeps the plane
on the stage

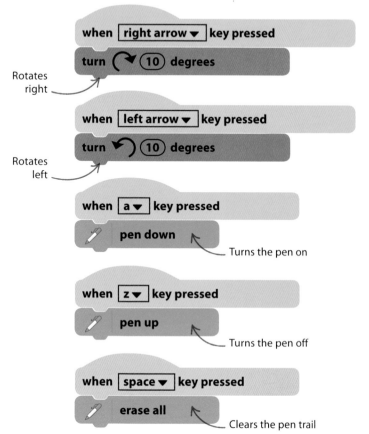

```
when right arrow ▼ key pressed
turn ↻ (10) degrees
```
Rotates
right

```
when left arrow ▼ key pressed
turn ↺ (10) degrees
```
Rotates
left

```
when a ▼ key pressed
🖊 pen down
```
Turns the pen on

```
when z ▼ key pressed
🖊 pen up
```
Turns the pen off

```
when space ▼ key pressed
🖊 erase all
```
Clears the pen trail

Turtle graphics

Using sprites to draw pictures
is called "turtle graphics".
That's because there's a type of
robot called a turtle that can be
moved around the floor to draw
pictures. The first programming
language to use turtle graphics
was called LOGO.

Variables

In coding, a variable is the name for a place where you can store information. They're used to remember things such as the score, a player's name, or a character's speed.

SEE ALSO

Maths **52–53 ⟩**

Variables **108–109 ⟩**
in Python

Creating a variable

You can create a variable to use in your program using the "Variables" section of the blocks palette. Once a variable has been created, new blocks appear in the blocks palette ready for you to use.

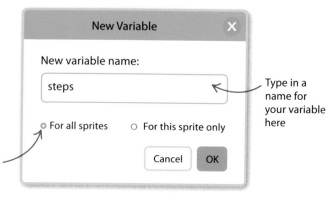

◁ **Storing data**
Variables are like boxes where you can store different bits of information for use in your program.

1 **Make a variable**
First, click the "Variables" button in the blocks palette. Then select the "Make a Variable" button.

Click here to create a variable

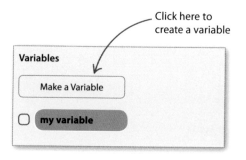

2 **Name the new variable**
Give the variable a name that will help you to remember what it does. Select which sprites will use the variable, then click "OK".

New Variable	✕
New variable name:	
steps	

Type in a name for your variable here

○ For all sprites   ○ For this sprite only

Cancel   OK

Choose whether the variable will be used by all sprites or just the one selected

3 **A new variable is created**
Once a new variable has been created, new blocks appear in the blocks palette. The menus inside these blocks let you select which variable they apply to, if you have created more than one.

Tick to show the variable on the stage

The variable block can be used inside other blocks

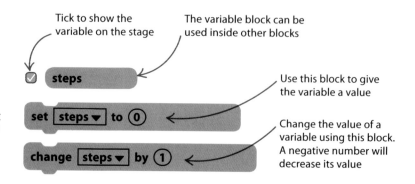

Use this block to give the variable a value

Change the value of a variable using this block. A negative number will decrease its value

Using a variable

Variables can be used to change a sprite's speed.
This simple bit of code shows you how.

1 **Set the value of a variable**
Create this code. Use the "set steps to 0" block
and change the number to 5. Drag the "move 10 steps"
block into the code, but drop the "steps" variable block
over the "10".

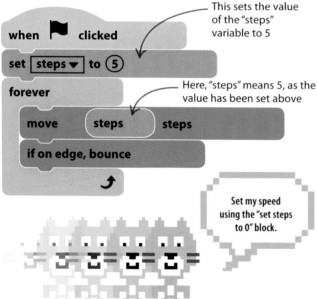

This sets the value
of the "steps"
variable to 5

Here, "steps" means 5, as the
value has been set above

Set my speed
using the "set steps
to 0" block.

2 **Changing the value of a variable**
Use the "change steps by 1" block to increase
the value of the variable "steps" by 1. Put it inside the
"forever" block, so the cat keeps on getting faster.

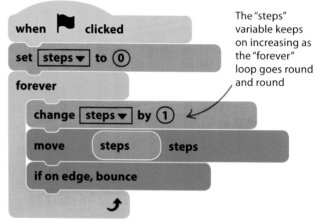

The "steps"
variable keeps
on increasing as
the "forever"
loop goes round
and round

Deleting variables

When you no longer want a variable,
right-click on it in the blocks palette and
then select "Delete the variable". You'll lose
any information that was in it.

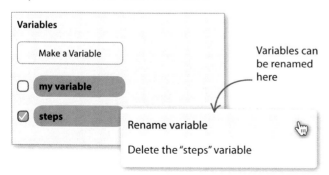

Variables can
be renamed
here

■ ■ EXPERT TIPS
Read-only variables

Some variables are set by Scratch and can't
be changed. They're still variables though,
because their values vary. These blocks are
known as sensing blocks.

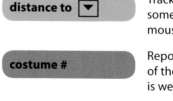

distance to ▼
Tracks the distance to
something, such as the
mouse-pointer.

costume #
Reports the number
of the costume a sprite
is wearing.

direction
Tells you which direction
a sprite is travelling in.

Maths

As well as storing numbers in variables (see pp.50–51), Scratch can be used to carry out all sorts of calculations using the "Operators" blocks.

SEE ALSO

❮ **50–51** Variables

Maths **112–113** ❯
in Python

Doing sums

There are four "Operators" blocks that can be used to do simple calculations. These are addition, subtraction, multiplication, and division.

△ **Addition**
The "+" block adds the two numbers in the block together.

△ **Subtraction**
The "–" block subtracts the second number from the first.

The "think" block is used here to print the result

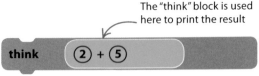

△ **Printing results**
Drag a "think" block into the code area and drop a "+" block inside it. Now add two numbers together and watch your sprite think the answer.

△ **Multiplication**
Computers use the "*" symbol for multiplication, because "x" looks like a letter.

△ **Division**
There's no division sign on the keyboard, so Scratch uses the "/" symbol instead.

Results in a variable

For more complex calculations, such as fixing the sale price of an item, instead of just using numbers you can use the value of a variable in a sum. The result can be stored in a variable too.

Variables are useful if you want to repeat the same sum with different values.

1 **Create variables**
Go to the "Variables" section of the blocks palette and create two variables – "sale price" and "price".

2 **Set the price**
Select the "set price" block and fix the price of an item to 50.

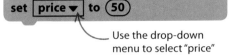

Use the drop-down menu to select "price"

3 **Calculate the sale price**
Use this code to calculate half the price of an item and set it as the sale price.

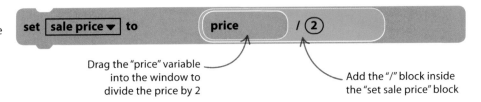

Drag the "price" variable into the window to divide the price by 2

Add the "/" block inside the "set sale price" block

Random numbers

The "pick random" block can be used to select a random number between two values. This block is useful for rolling dice in a game or for when you want to mix up a sprite's costumes.

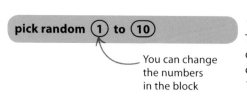

◁ **Pick a random number**
To pick a random month, change the numbers to choose a number between 1 and 12.

You can change the numbers in the block

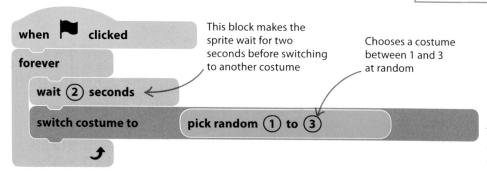

This block makes the sprite wait for two seconds before switching to another costume

Chooses a costume between 1 and 3 at random

◁ **Switching costumes**
This code changes a sprite's costume at random every two seconds.

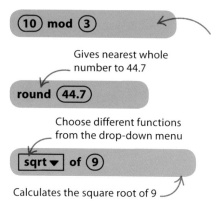

◁ **Random costumes**
Costumes can make a sprite appear to move its body, or might give it different clothes, as shown here.

Hard maths

Simple "Operators" blocks can do most calculations, but Scratch can also do more complex maths. The "mod" block divides two numbers and gives the remainder, which is the number that is left over. The "round" block rounds to the nearest whole number, and the "sqrt" block gives the square root of a number.

Divides 10 by 3 and gives the remainder – the number that is left over

Gives nearest whole number to 44.7

Choose different functions from the drop-down menu

◁ **More maths**
The "Operators" section has blocks of advanced maths functions that can be used to do complex calculations.

Calculates the square root of 9

Strings and lists

In programming, a sequence of letters and symbols is called a "string". Strings can contain any character on the keyboard (including spaces) and be of any length. Strings can also be grouped together in lists.

SEE ALSO

❮ **50–51** Variables

Strings **114–115** ❯
in Python

Keyboard characters are lined up as if they were hanging from a string

Working with words

Programs often need to remember words, such as a player's name. Variables can be created to remember these words. Scratch programs can also ask the user questions, which they answer by typing into a text box that pops up. The following code asks for the user's name, and then makes a sprite say "Hello" to them.

1 **Create a new variable**
Click the "Variables" button in the blocks palette and click the "Make a Variable" button. Create a variable called "greeting".

Variables

Make a Variable

☑ greeting

☐ my variable

Name your variable "greeting"

2 **Asking a question**
This code makes the sprite ask a question. Whatever the user types into the text box that pops up on the screen is stored in a new variable called "answer". The code then combines the strings contained in the "greeting" and "answer" variables to greet the user.

This block puts "Hello " into the variable "greeting". Leave a space at the end of "Hello " to make the output of the program neater

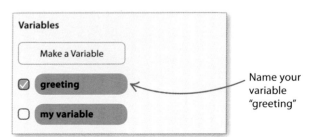

The "ask" box (from the "Sensing" section of the blocks palette) makes a text box appear, which the user types their answer into

The "answer" variable (from the "Sensing" section) contains whatever the user typed into the text box

The "say" bubble creates a speech bubble for the sprite

The "greeting" variable holds the string "Hello "

Making lists

Variables are perfect if you just want to remember one thing. To remember lots of similar things, lists can be used instead. Lists can store many items of data (numbers and strings) at the same time – for example, all of the high scores in a game. The following program shows one way of using a list.

1 Create a list
Start a new project. Go into the "Variables" section of the blocks palette and click the "Make a List" button. Give your list the name "sentence".

Call your list "sentence"

2 Using your list
This code asks the user to type words into a list. Each word appears in the sprite's speech bubble as it is added to the list.

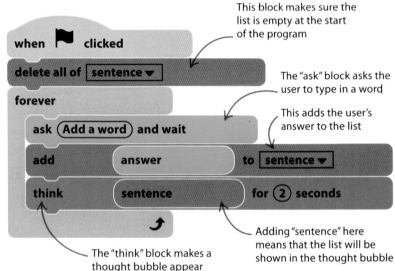

This block makes sure the list is empty at the start of the program

The "ask" block asks the user to type in a word

This adds the user's answer to the list

Adding "sentence" here means that the list will be shown in the thought bubble

The "think" block makes a thought bubble appear

3 Seeing the list
If you tick the box beside the list in the blocks palette, the list is shown on the stage. You can see each new word as it's added to the list.

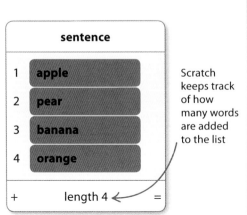

Scratch keeps track of how many words are added to the list

· · EXPERT TIPS
Playing with lists

These blocks can be used to change the contents of a list. Each item in a list has a number – the first item is number 1, and so on. These numbers can be used to remove, insert, or replace items.

Deletes the first item in the list

Adds "cherry" as the first item in the list

Replaces the first list item with "cherry"

Co-ordinates

To put a sprite in a particular spot, or to find out its exact location, you can use co-ordinates. Co-ordinates are a pair of numbers that pinpoint a sprite's position on the stage using an x and y grid.

SEE ALSO

❮ **38–39** Making things move

❮ **52–53** Maths

x and y positions

The x and y positions of a sprite are shown on the Scratch interface. It can be helpful to know a sprite's co-ordinates when writing code.

Type in these boxes to change the sprite's current co-ordinates

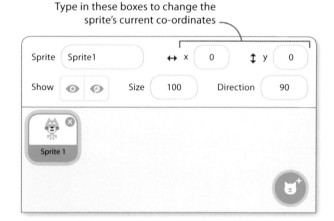

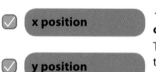

◁ **Show co-ordinates on the stage**

Tick the boxes beside the "x position" and "y position" blocks to show a sprite's position on the stage.

△ **Position of a sprite**

You can see a sprite's current co-ordinates in the information panel above the sprites list.

x and y grid

To pinpoint a spot, count the number of steps left or right, and up or down, from the middle of the stage. Steps to the left or right are called "x". Steps up or down are called "y". Use negative numbers to move left and down.

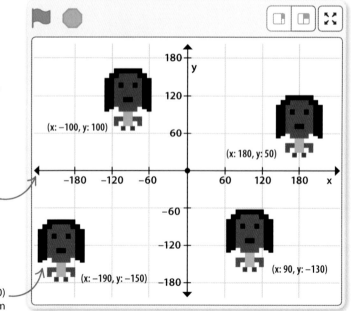

The stage is based upon an x and y grid

This sprite is 190 steps left (−190) and 150 steps down (−150) from the middle of the stage

Moving the sprite

Co-ordinates are used to move a sprite to a particular spot on the stage. It doesn't matter how near or far away the spot is. The "glide 1 secs to x:0 y:0" block from the "Motion" section of the blocks palette makes the sprite glide there smoothly.

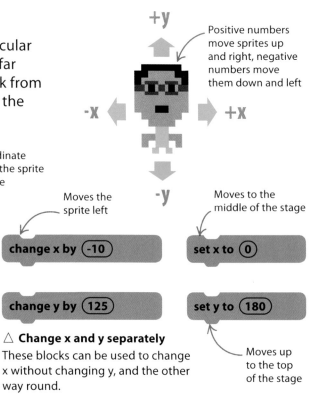

Positive numbers move sprites up and right, negative numbers move them down and left

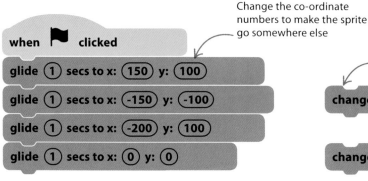

Change the co-ordinate numbers to make the sprite go somewhere else

```
when 🏳 clicked
glide 1 secs to x: 150 y: 100
glide 1 secs to x: -150 y: -100
glide 1 secs to x: -200 y: 100
glide 1 secs to x: 0 y: 0
```

△ **Control the sprite with code**
Can you work out the path the sprite will take when you run this code? Try it and see!

Moves the sprite left

```
change x by -10

change y by 125
```

Moves to the middle of the stage

```
set x to 0

set y to 180
```

Moves up to the top of the stage

△ **Change x and y separately**
These blocks can be used to change x without changing y, and the other way round.

Crazy horse's trip

Try this fun code to test out co-ordinates. Select the "Horse" sprite from the sprites library and give it the code shown below. This program uses the "go to x:0 y:0" block to keep moving the horse to a random position, drawing a line behind it as it goes.

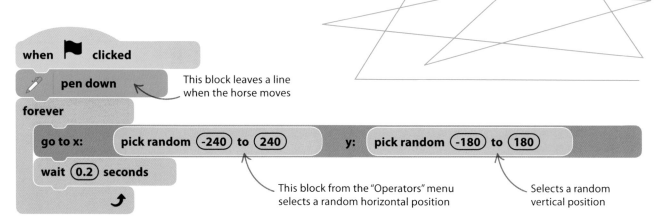

```
when 🏳 clicked
🖊 pen down
forever
    go to x: pick random -240 to 240   y: pick random -180 to 180
    wait 0.2 seconds
```

This block leaves a line when the horse moves

This block from the "Operators" menu selects a random horizontal position

Selects a random vertical position

Make some noise!

Scratch programs don't have to be silent. Use the pink "Sound" blocks to try out sound effects and create music. You can also use sound files you already have or record brand new sounds for your program.

SEE ALSO

Sensing **66–67 〉**
and detecting

Monkey **74–81 〉**
mayhem

Adding sounds to sprites

To play a sound, it must be added to a sprite. Each sprite has its own set of sounds. To control them, click the "Sounds" tab above the blocks palette.

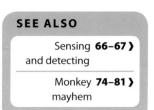

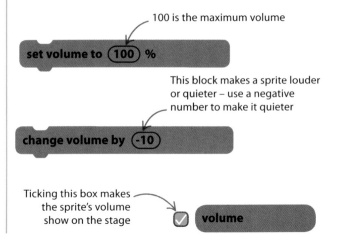

Upload a recording from the computer

Record a sound using the computer's microphone

Click here to select a sound effect from Scratch's library

Choose a Sound

Playing a sound

There are two blocks that play sounds: "start sound" and "play sound until done". "Until done" makes the program wait until the sound has finished before it moves on.

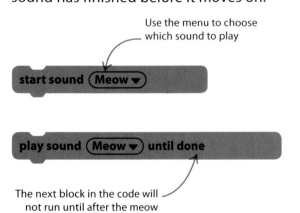

Use the menu to choose which sound to play

start sound (Meow ▼)

play sound (Meow ▼) until done

The next block in the code will not run until after the meow sound has finished playing

Turn up the volume

Each sprite has its own volume control, which is set using numbers. 0 is silent and 100 is the loudest.

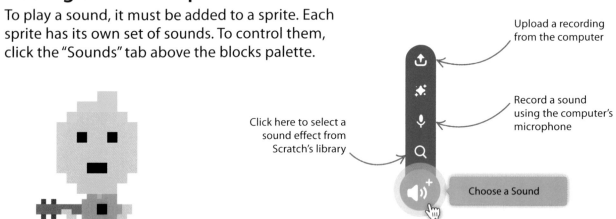

100 is the maximum volume

set volume to (100) %

This block makes a sprite louder or quieter – use a negative number to make it quieter

change volume by (-10)

Ticking this box makes the sprite's volume show on the stage

✓ volume

Making your own music

Scratch has blocks that can be used to invent musical sounds. You have a whole orchestra of instruments to conduct, as well as a full drum kit. The length of each note is measured in beats. To use the music blocks, click "Add Extension" at the bottom left then choose "Music".

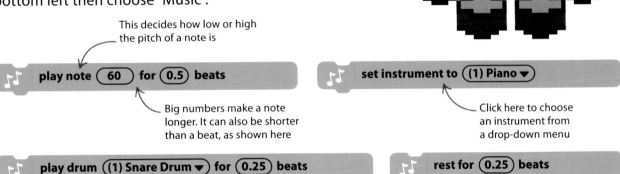

This decides how low or high the pitch of a note is

`play note (60) for (0.5) beats`

Big numbers make a note longer. It can also be shorter than a beat, as shown here

`set instrument to ((1) Piano ▼)`

Click here to choose an instrument from a drop-down menu

`play drum ((1) Snare Drum ▼) for (0.25) beats`

Use this menu to choose between different types of drum

`rest for (0.25) beats`

This block adds a silent break in the music. Higher numbers will give you a longer break

Playing music

Connecting notes together makes a tune. Create a new variable called "note" (see pages 50–51), and then add the following code to any sprite to create a piece of music.

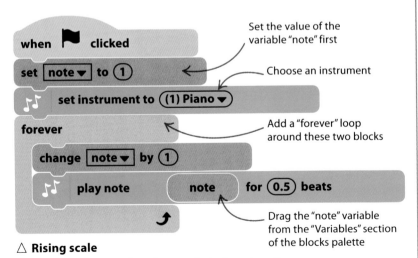

Set the value of the variable "note" first

Choose an instrument

Add a "forever" loop around these two blocks

Drag the "note" variable from the "Variables" section of the blocks palette

△ **Rising scale**

This code makes a series of notes that play when the green flag is clicked. The pitch of each note gets higher one step at a time, and each note plays for half a beat.

▪▪ EXPERT TIPS

Tempo

The speed of music is called its tempo. The tempo decides how long a beat is within a piece of music. There are three blocks for managing the tempo.

`set tempo to (60)`

The tempo is measured in beats per minute, or "bpm".

`change tempo by (20)`

Increase the tempo to make your music faster, or use a negative number to make it slower.

`☑ ♫ tempo`

Ticking this box makes the sprite's tempo show on the stage.

> PROJECT 2

Roll the dice

Simple programs can be both useful and fun. This program creates a dice that can be rolled. Play it to see who can get the highest number, or use it instead of a real dice when you play a board game.

SEE ALSO

‹ **40–41** Costumes

‹ **46–47** Simple loops

‹ **50–51** Variables

‹ **52–53** Maths

How to create a rolling dice

The dice in this program uses six costumes. Each costume shows a face of the dice with a different number on it – from one to six.

1 Select the "Paint" icon in the "Choose a Sprite" menu to draw a new sprite.

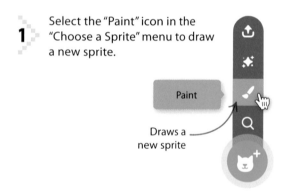

Draws a new sprite

2 Select "Convert to Bitmap". Click the rectangle button on the left of the painting area. To make your dice colourful, select a solid colour from the palette (see box right). Then in the painting area hold down the "shift" key, press the left mouse button, and then drag the mouse-pointer to make a square in the middle.

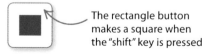

The rectangle button makes a square when the "shift" key is pressed

· · EXPERT TIPS

Changing colours

Above the painting area are the colour controls. Click the solid rectangle to draw a block of solid colour. Click the empty rectangle to draw an outline of a square or rectangle. Use the spinner buttons to change the thickness of the square's lines. To choose a colour, click on the "Fill" tab. A colour palette will appear in the drop-down menu.

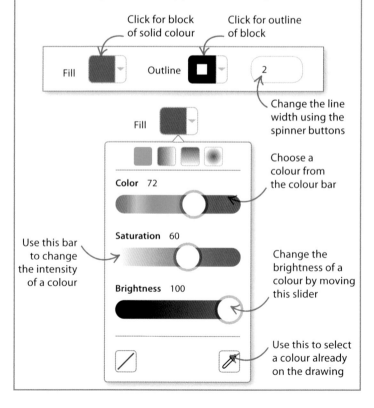

Click for block of solid colour

Click for outline of block

Fill

Outline

2

Change the line width using the spinner buttons

Fill

Color 72

Choose a colour from the colour bar

Saturation 60

Use this bar to change the intensity of a colour

Brightness 100

Change the brightness of a colour by moving this slider

Use this to select a colour already on the drawing

3 Right-click on your costume to the left of the painting area, and choose "duplicate". Repeat this step until you have six costumes.

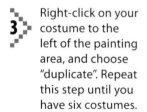

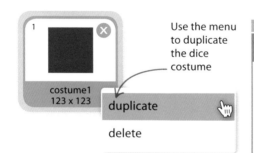

Use the menu to duplicate the dice costume

Rotation tool

To make the dice appear to roll when the code is run, you can rotate each costume to a different angle. Click on the "Convert to Vector" button in the bottom left-hand corner and select the "Select" tool. When you click back on to the painting area, a rotation tool will appear.

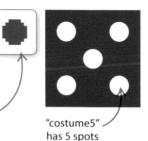

Click and drag this control to rotate the dice

4 Select a costume. Click the circle button on the painting area and choose a solid white colour from the palette. Add spots to each of the six costumes until you have made all six sides of a dice.

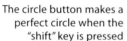

The circle button makes a perfect circle when the "shift" key is pressed

"costume5" has 5 spots

5 Add the code below to the dice sprite. Press the spacebar to roll the dice. Try it a few times to check you can see all of the costumes.

Clicking the spacebar rolls the dice

```
when  space ▼  key pressed
switch costume to    pick random ① to ⑥
```

This block selects a random costume

6 Sometimes you'll roll the same number twice, and it looks like the program isn't working because the image doesn't change. This code makes the dice change costumes five times before it stops. Each time you press the spacebar, it looks like it's rolling.

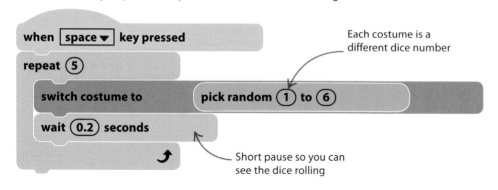

Each costume is a different dice number

Short pause so you can see the dice rolling

Don't forget to save your work

True or false?

Computers decide what to do by asking questions and determining whether the answers are true or false. Questions that only have two possible answers are called "Boolean expressions".

SEE ALSO

Decisions and **64–65 〉**
branches

Making **118–119 〉**
decisions

Comparing numbers

You can compare numbers using the "=" block from the "Operators" section of the blocks palette.

◁ **The "=" block**
This block will give one of two answers – "true" if the two numbers in the boxes are equal, and "false" if they aren't.

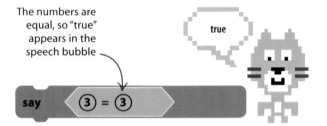

The numbers are equal, so "true" appears in the speech bubble

△ **True answer**
Using an "=" block inside a speech block will make "true" or "false" appear in a sprite's speech bubble.

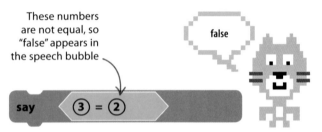

These numbers are not equal, so "false" appears in the speech bubble

△ **False answer**
If the numbers in the block are different, the sprite's speech bubble will contain the word "false".

Comparing variables

You can use variables inside comparison blocks. It's not worth comparing fixed numbers because the result will always be the same, whereas the value of variables can change.

△ **Create a variable**
Click the "Variables" button in the blocks palette and create a new variable called "age". Set its value to 10 (click on the block to make sure the value has changed). Drag the "age" variable into the comparison blocks.

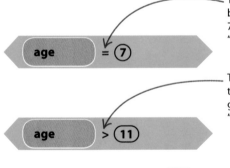

This sign means "equals", so the block is asking if "age" is equal to 7. The answer here is "false", as "age" is 10

This sign means "more than", so the block is asking if "age" is greater than 11. The answer is "false", as 10 is not bigger than 11

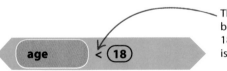

This sign means "less than", so the block is asking if "age" is less than 18. The answer will be "true", as 10 is smaller than 18

△ **Comparing numbers**
Find the green comparison blocks in the "Operators" menu. As well as checking whether two numbers are equal, you can check whether one is higher or lower than another.

Comparing words

The "=" block is not just used for numbers – it can also be used to check whether two strings are the same. It ignores capital letters when comparing strings.

△ **Create a variable**
To experiment with comparing strings, create a new variable called "name" and set its value to "Lizzie".

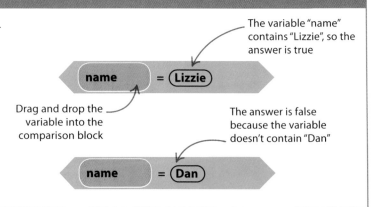

The variable "name" contains "Lizzie", so the answer is true

Drag and drop the variable into the comparison block

The answer is false because the variable doesn't contain "Dan"

Not!

The "not" block can simplify things by reversing the answer of a Boolean expression. For example, it's easier to check if someone's age is not 10 than to check every other possible age.

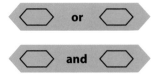

◁ **The "not" block**
The "not" block changes the answer around, from true to false and from false to true.

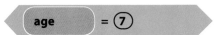

△ **Without the "not" block**
Here, 10 isn't equal to 7, so the answer is false.

△ **With the "not" block**
Adding the "not" block to the same question changes the answer. As 7 does not equal 10, the answer is now true.

Combining questions

To ask more complicated questions, you can combine comparison blocks and ask more than one question at the same time.

△ **Comparison blocks**
The "or" and "and" blocks are used to combine Boolean expressions in different ways.

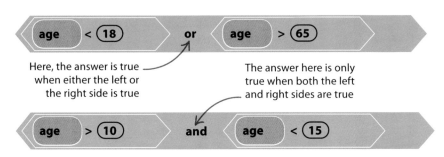

Here, the answer is true when either the left or the right side is true

The answer here is only true when both the left and right sides are true

◁ **In practice**
The top block checks whether someone is younger than 18 or older than 65. The bottom block checks if they are aged 11, 12, 13, or 14.

Decisions and branches

By testing whether something is true or false you can use this information to tell the computer what to do next. It will perform a different action depending on whether the answer is true or false.

SEE ALSO

⟨ **62–63** True or false?

Sensing and **66–67** ⟩ detecting

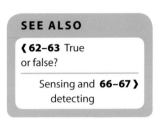

Making decisions

The "if" blocks use Boolean expressions to decide what to do next. To use them, put other blocks inside their "jaws". The blocks inside the "if" blocks will only run if the answer to the Boolean expression is true.

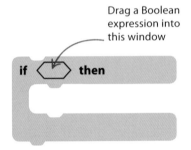

Drag a Boolean expression into this window

△ **"if-then" block**

If a Boolean expression is true, the blocks between the "if-then" block's jaws will run.

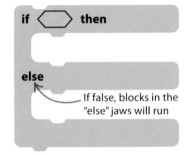

△ **"if-then-else" block**

If the Boolean expression is true, the first set of blocks runs. If not, the second set runs.

If false, blocks in the "else" jaws will run

Using the "if-then" block

The "if-then" block lets you choose whether or not to run a specific bit of code depending on the answer to a Boolean expression. Attach this code to the cat sprite to try it out.

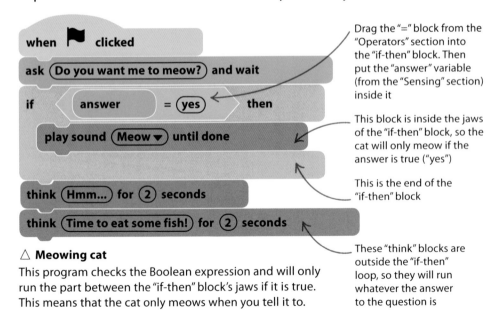

when ▶ clicked

ask (Do you want me to meow?) and wait

if ⟨ answer = (yes) ⟩ then

   play sound (Meow ▼) until done

think (Hmm...) for ② seconds

think (Time to eat some fish!) for ② seconds

Drag the "=" block from the "Operators" section into the "if-then" block. Then put the "answer" variable (from the "Sensing" section) inside it

This block is inside the jaws of the "if-then" block, so the cat will only meow if the answer is true ("yes")

This is the end of the "if-then" block

These "think" blocks are outside the "if-then" loop, so they will run whatever the answer to the question is

△ **Meowing cat**

This program checks the Boolean expression and will only run the part between the "if-then" block's jaws if it is true. This means that the cat only meows when you tell it to.

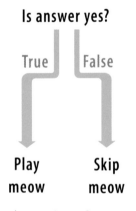

Is answer yes?

True   False

Play meow   Skip meow

△ **How it works**

The program checks whether the Boolean expression is true. If it is, it runs the blocks inside the "if-then" block's jaws.

Branching instructions

Often you want a program to do one thing if a condition is true, and something else if it is not. The "if-then-else" block gives a program two possible routes, called "branches". Only one branch will run, depending on the answer to the Boolean expression.

▽ **Branching program**
This program has two branches: one will run if the answer is "yes", and the other will run if it is not.

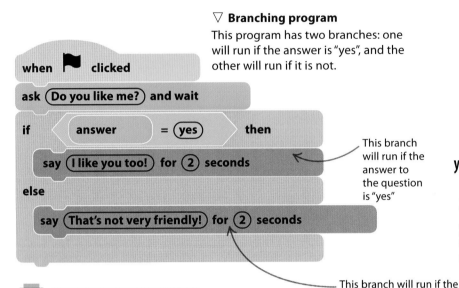

This branch will run if the answer to the question is "yes"

This branch will run if the answer to the question is anything except "yes"

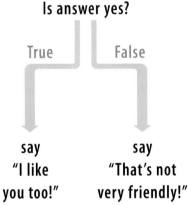

Is answer yes?

True   False

say "I like you too!"   say "That's not very friendly!"

△ **How it works**
The program checks whether you typed in "yes". If so, it shows the first message. If not, it shows the second.

Boolean shapes

The Boolean expression blocks in Scratch have pointed ends. You can put them into some non-pointed shaped holes too.

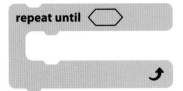

mouse down?

△ **"Sensing" blocks**
These blocks can test whether a sprite is touching another sprite, or whether a button is pressed.

repeat until ⬡

△ **"Control" blocks**
Several "Control" blocks have Boolean-shaped holes in them for Boolean expressions.

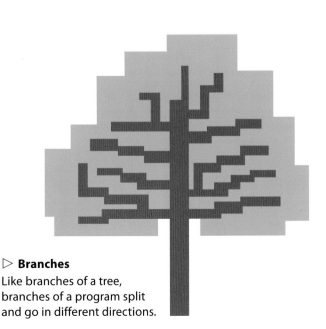

▷ **Branches**
Like branches of a tree, branches of a program split and go in different directions.

Sensing and detecting

SEE ALSO

❮ **40–41** Costumes

❮ **56–57** Co-ordinates

The "Sensing" blocks enable your program to see what is happening on your computer. They can detect keyboard controls, and let sprites react when they touch each other.

This block checks if a key is being pressed. You can choose which key to check for

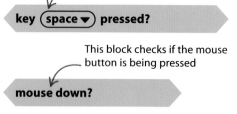

This block checks if the mouse button is being pressed

Keyboard controls

Using "Sensing" blocks with "if-then" blocks allows you to move a sprite around the screen using the keyboard. The "key pressed?" block has a menu of most of the keys on the keyboard, so a sprite can be programmed to react to almost any key. You can also link actions to the click of a mouse button.

△ **"Sensing" blocks**

Adding these blocks into an "if-then" block allows the program to detect if a mouse button or key is being pressed.

Putting everything inside a "forever" block means the code repeatedly checks for key presses

The code checks to see if the up arrow is pressed. If it is, the sprite moves upwards on the screen

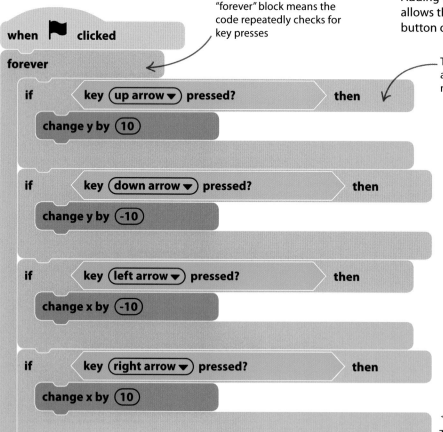

△ **Controlling sprites**

Keyboard controls give you precise control over your sprites, which is especially useful in games.

◁ **Movement code**

This code lets you move sprites up, down, left, or right using the arrow keys on the keyboard.

Sprite collisions

It can be useful to know when one sprite touches another – in games, for example. Use "Sensing" blocks to make things happen when sprites touch each other, or when a sprite crosses an area that is a certain colour.

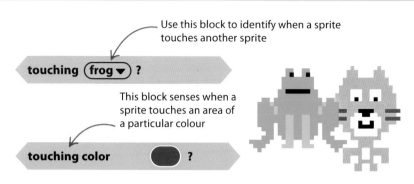

Use this block to identify when a sprite touches another sprite

touching (frog ▼) ?

This block senses when a sprite touches an area of a particular colour

touching color ⬤ ?

Using "Sensing" blocks

Use the "Sensing" blocks to turn your controllable cat into a game. Start by adding the movement code created on the opposite page to the cat sprite, then add the "Room 1" backdrop and the elephant sprite. Using the "Sounds" tab, add the "Trumpet2" sound effect to the elephant, then build it the code below.

▽ **Find the elephant**
This code uses "Sensing" blocks to control the relationship between the cat and the elephant. As the cat gets nearer, the elephant grows. When the cat touches it, the elephant switches costume, makes a sound, and hides somewhere else.

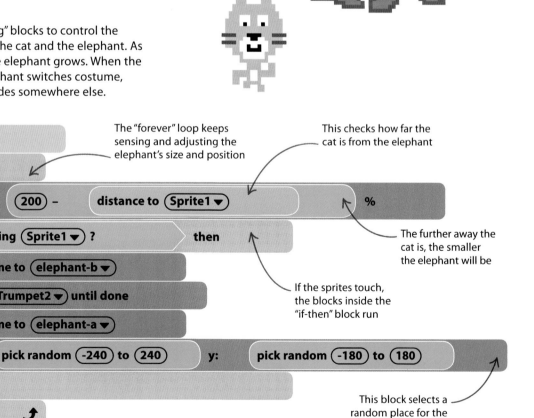

The "forever" loop keeps sensing and adjusting the elephant's size and position

This checks how far the cat is from the elephant

```
when ⚑ clicked
forever
    set size to (200) – (distance to (Sprite1 ▼)) %
    if < touching (Sprite1 ▼) ? > then
        switch costume to (elephant-b ▼)
        play sound (Trumpet2 ▼) until done
        switch costume to (elephant-a ▼)
        go to x: (pick random (-240) to (240)) y: (pick random (-180) to (180))
```

The further away the cat is, the smaller the elephant will be

If the sprites touch, the blocks inside the "if-then" block run

This block selects a random place for the elephant to hide

Complex loops

Simple loops are used to repeat parts of a program forever, or a certain number of times. Other, cleverer loops can be used to write programs that decide exactly when to repeat instructions.

SEE ALSO

❰ **46–47** Simple loops

❰ **62–63** True or false?

Looping until something happens

Add the "Dog1" sprite to a project, and then give the code below to the cat sprite. When you run the code, the "repeat until" block makes sure the cat keeps moving until it touches the dog. It will then stop and say "Ouch!"

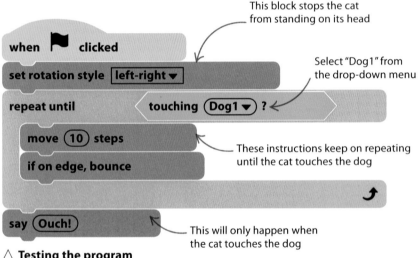

This block stops the cat from standing on its head

Select "Dog1" from the drop-down menu

These instructions keep on repeating until the cat touches the dog

This will only happen when the cat touches the dog

△ **Testing the program**
Move the dog out of the cat's way and run the program. Then drag and drop the dog into the cat's path to see what happens.

△ **"Repeat until" block**
The blocks inside the "repeat until" block keep repeating until the condition is true (the cat touches the dog).

Ouch!

Stop!

Another useful "Control" block is the "stop all" block, which can stop code from running. It's useful if you want to stop sprites moving at the end of a game.

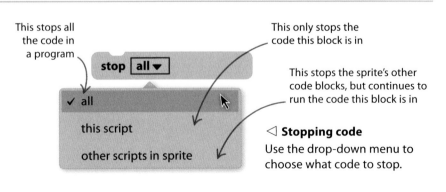

This stops all the code in a program

This only stops the code this block is in

This stops the sprite's other code blocks, but continues to run the code this block is in

◁ **Stopping code**
Use the drop-down menu to choose what code to stop.

Waiting

It's easier to play a game or see what's going on in a program if you can make the code pause for a moment. Different blocks can make the code wait a number of seconds or until something is true.

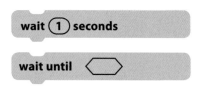

◁ **Waiting blocks**
The "wait seconds" block waits a set amount of time. The "wait until" block responds to what's happening in the program.

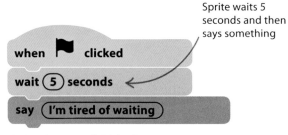

Sprite waits 5 seconds and then says something

△ **"wait seconds" block**
With the "wait seconds" block you can enter the number of seconds you want a sprite to wait.

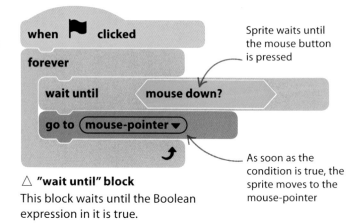

Sprite waits until the mouse button is pressed

As soon as the condition is true, the sprite moves to the mouse-pointer

△ **"wait until" block**
This block waits until the Boolean expression in it is true.

Magnetic mouse

Different loops can be used together to make programs. This program starts once the mouse button is pressed. The sprite follows the mouse-pointer until the mouse button is released. It then jumps up and down five times. The whole thing then repeats itself because it's all inside a "forever" loop.

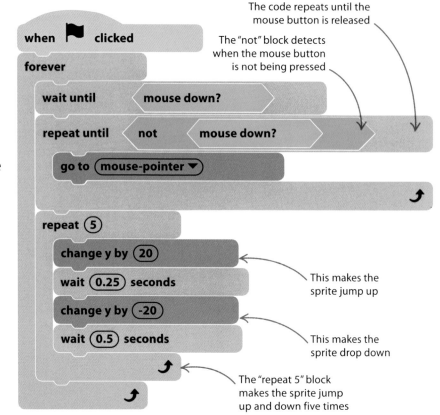

The code repeats until the mouse button is released

The "not" block detects when the mouse button is not being pressed

This makes the sprite jump up

This makes the sprite drop down

The "repeat 5" block makes the sprite jump up and down five times

▷ **Nested loops**
Pay careful attention to how the loops are nested inside the "forever" block.

Sending messages

Sometimes it's useful for sprites to communicate with each other. Sprites can use messages to tell other sprites what to do. Scratch also lets you create conversations between sprites.

SEE ALSO

❮ 38–39 Making things move

❮ 40–41 Costumes

❮ 44–45 Events

Broadcasting

The broadcast blocks in the "Events" menu enable sprites to send and receive messages. Messages don't contain any information other than a name, but can be used to fine-tune a sprite's actions. Sprites only react to messages that they are programmed to respond to – they ignore any other messages.

This "Events" block lets a sprite send a message to all the other sprites

`broadcast (message1 ▼)`

This block starts the code when a sprite receives a message

`when I receive (message1 ▼)`

△ **Broadcast blocks**
One type of broadcast block lets a sprite send a message. The other tells the sprite to receive a message. Choose an existing message or create a new one.

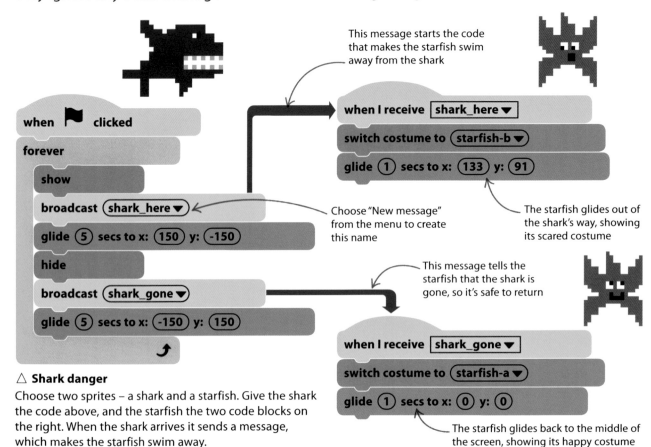

This message starts the code that makes the starfish swim away from the shark

Choose "New message" from the menu to create this name

The starfish glides out of the shark's way, showing its scared costume

This message tells the starfish that the shark is gone, so it's safe to return

The starfish glides back to the middle of the screen, showing its happy costume

△ **Shark danger**
Choose two sprites – a shark and a starfish. Give the shark the code above, and the starfish the two code blocks on the right. When the shark arrives it sends a message, which makes the starfish swim away.

Conversations

To create a conversation between sprites use "broadcast message and wait" blocks with "say" blocks, which make your sprites talk using speech bubbles. Start a new project and add two monkey sprites to it. Give the code on the left to one monkey, and the two code blocks on the right to the other.

broadcast (message1 ▼) and wait

△ **Waiting blocks**
This block sends a message, then waits for all the code that reacts to the message to finish before the program continues.

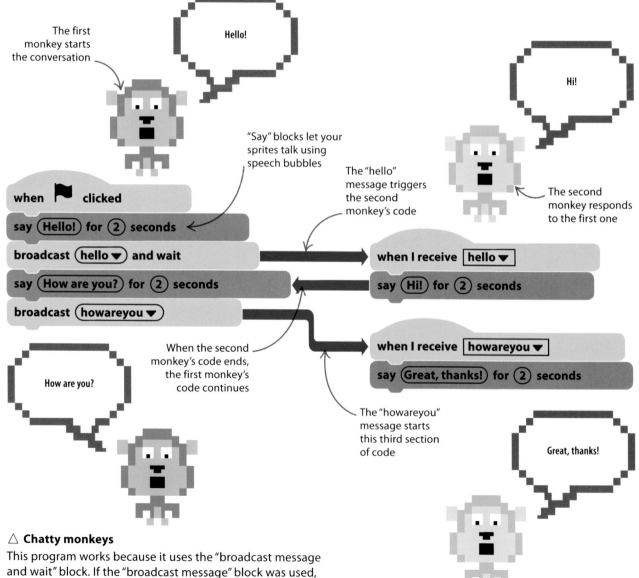

The first monkey starts the conversation

Hello!

Hi!

"Say" blocks let your sprites talk using speech bubbles

The "hello" message triggers the second monkey's code

The second monkey responds to the first one

```
when 🏳 clicked
say (Hello!) for (2) seconds
broadcast (hello ▼) and wait
say (How are you?) for (2) seconds
broadcast (howareyou ▼)
```

```
when I receive (hello ▼)
say (Hi!) for (2) seconds
```

When the second monkey's code ends, the first monkey's code continues

```
when I receive (howareyou ▼)
say (Great, thanks!) for (2) seconds
```

How are you?

The "howareyou" message starts this third section of code

Great, thanks!

△ **Chatty monkeys**
This program works because it uses the "broadcast message and wait" block. If the "broadcast message" block was used, the monkeys would talk over each other.

Creating blocks

To avoid repeating the same set of blocks over and over again, it's possible to take a shortcut by creating new blocks. Each new block can contain several different instructions.

SEE ALSO

⟨ **50–51** Variables

Time to **82–83** ⟩
experiment

Making your own block

You can make your own blocks in Scratch that run code when they're used. Try this example to see how they work. Programmers call these reusable pieces of code "subprograms" or "functions".

1 Create a new block
Click on the "My Blocks" button, and then select "Make a Block". Type the word "jump" and click "OK".

2 New block appears
Your new block "jump" appears in the blocks palette, and a "define" block appears in the code area.

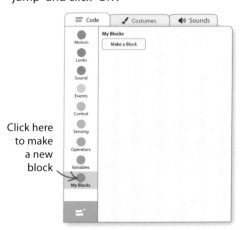

Click here to make a new block

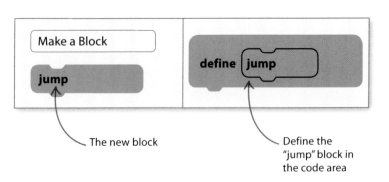

Make a Block

jump

define jump

The new block

Define the "jump" block in the code area

3 Define the block
The "define" block tells Scratch which blocks to run when using the new block. Add this code to define the block.

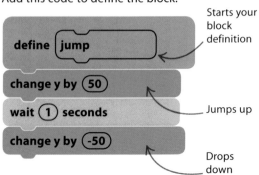

define jump

change y by (50)

wait (1) seconds

change y by (-50)

Starts your block definition

Jumps up

Drops down

4 Use the block in the code
The new block can now be used in the code. It's as if those jumping blocks were in the code individually.

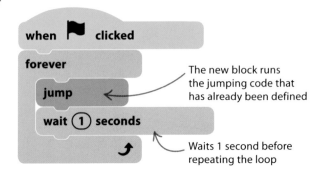

when 🏳 clicked

forever

jump

wait (1) seconds

The new block runs the jumping code that has already been defined

Waits 1 second before repeating the loop

Blocks with inputs

Windows in a new block can be used to give it numbers and words to work with. These holes can be used to change how far the block moves a sprite.

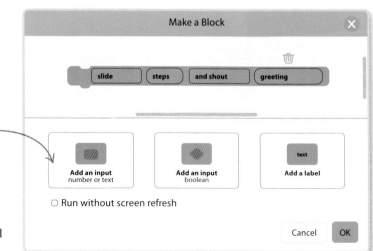

Click here to add a number or a string

1 **Make a new block**
Make a new block called "slide". Now select "Add an input number or text" and type "steps". Select "Add a label" and change it to "and shout". Click "Add an input number or text" again and call it "greeting". Then click "OK".

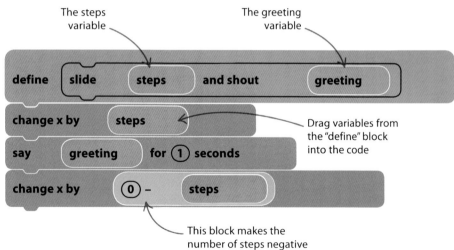

The steps variable

The greeting variable

2 **Define the block**
In the "define" block, the holes are replaced with variables called "steps" and "greeting". Drag these variables from the "define" block into the code wherever you need them. Add this code to your sprite.

Drag variables from the "define" block into the code

This block makes the number of steps negative

3 **Use the block in the code**
Now add the below code to a sprite. By putting different numbers of steps and greetings into the block, you can make your sprite behave differently.

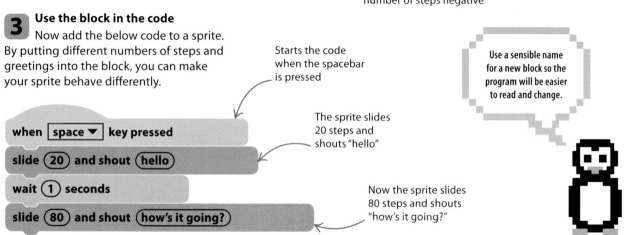

Starts the code when the spacebar is pressed

The sprite slides 20 steps and shouts "hello"

Now the sprite slides 80 steps and shouts "how's it going?"

Use a sensible name for a new block so the program will be easier to read and change.

► PROJECT 3

Monkey mayhem

This exciting, fast-paced game brings together all of the Scratch skills you've learned so far. Follow these steps to create your very own "Monkey mayhem" and see if you can hit the bat with the bananas!

SEE ALSO

❰ **40–41** Costumes

❰ **38–39** Making things move

❰ **66–67** Sensing and detecting

Getting started

Start a new Scratch project. The cat sprite isn't needed for this project. To remove it, right-click on it in the sprites list and then click "delete" in the menu. This will leave you a blank project to work on.

■ ■ EXPERT TIPS

Avoiding errors

This is the biggest Scratch program you've tried so far, so you might find that the game doesn't always work as you expect it to. Here are some tips to help things run smoothly:

Make sure you add the code to the correct sprite.

Follow the instructions carefully. Remember to make a variable before using it.

Check that all the numbers in the blocks are correct.

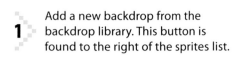

1 Add a new backdrop from the backdrop library. This button is found to the right of the sprites list.

Click here to add a new backdrop from the backdrop library

Choose a Backdrop

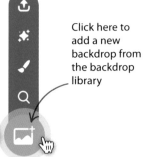

2 Search for the "Wall 1" backdrop and select it. The brick wall works well for this game, but if you prefer, you could use a different backdrop instead.

Click on a backdrop in the backdrop library to make it appear on the stage

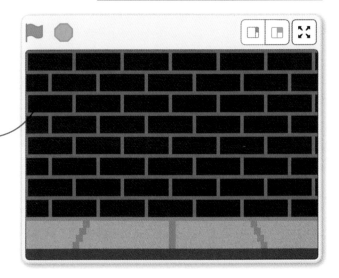

3 Go to the sprites library to add a new sprite to the game. Search for "Monkey" and select it. The user will control this sprite in the game.

Monkey

Click here to choose a new sprite from the library

Choose a Sprite

4 Give the monkey the code below. Remember – all of the different blocks can be found in the blocks palette, organized by colour. In this code, "Sensing" blocks are used to move the monkey around the stage using the keyboard arrow keys. Run the code when you've finished to check it works.

The arrow keys on the keyboard will make the monkey run left and right

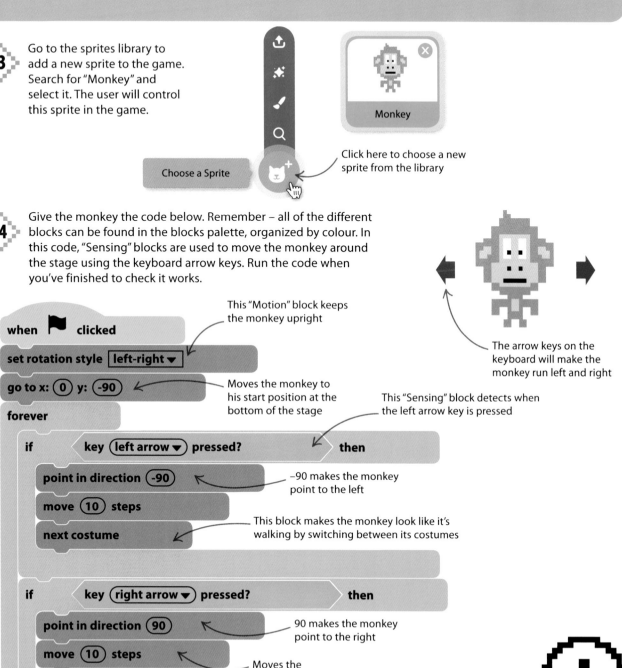

This "Motion" block keeps the monkey upright

when ⚑ clicked

set rotation style left-right ▼

go to x: 0 y: -90

Moves the monkey to his start position at the bottom of the stage

forever

if key left arrow ▼ pressed? then

This "Sensing" block detects when the left arrow key is pressed

point in direction -90

–90 makes the monkey point to the left

move 10 steps

next costume

This block makes the monkey look like it's walking by switching between its costumes

if key right arrow ▼ pressed? then

point in direction 90

90 makes the monkey point to the right

move 10 steps

Moves the monkey 10 steps

next costume

Don't forget to save your work

⊙ MONKEY MAYHEM

Adding more sprites

The monkey can now be moved across the stage using the left and right arrow keys. To make the game more interesting, add some more sprites. Give the monkey some bananas to throw, and a bat to throw them at!

5 Add the "Bananas" sprite from the sprites library, then give it this code. When the game starts, the monkey will be holding the bananas. When the spacebar is pressed, they will shoot vertically up the stage. The bananas then reappear at one side of the stage, where they can be picked up again.

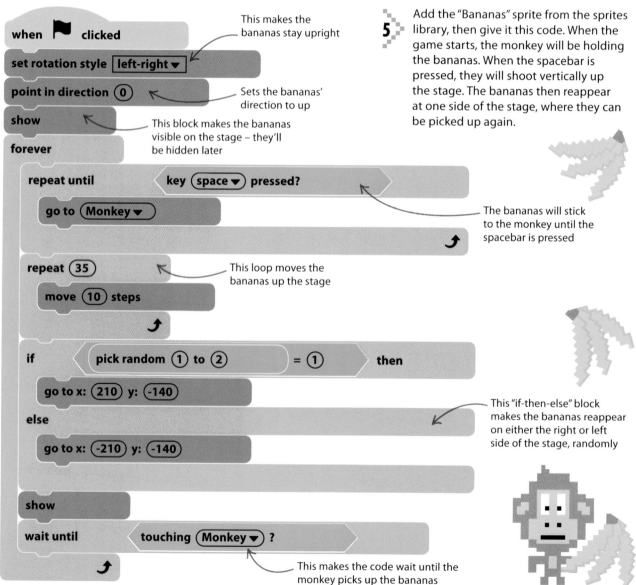

```
when 🏳 clicked
set rotation style [left-right ▼]
point in direction (0)
show
forever
    repeat until      key (space ▼) pressed?
        go to (Monkey ▼)

    repeat (35)
        move (10) steps

    if      pick random (1) to (2)  = (1)    then
        go to x: (210) y: (-140)
    else
        go to x: (-210) y: (-140)

    show
    wait until      touching (Monkey ▼) ?
```

This makes the bananas stay upright

Sets the bananas' direction to up

This block makes the bananas visible on the stage – they'll be hidden later

The bananas will stick to the monkey until the spacebar is pressed

This loop moves the bananas up the stage

This "if-then-else" block makes the bananas reappear on either the right or left side of the stage, randomly

This makes the code wait until the monkey picks up the bananas

6 The next step is to add a flying bat and make it drop to the ground if it's hit by the bananas. Add "Bat" from the sprites library, then create a new variable called "Speed" (for the bat sprite only). To create a new variable, first click the "Variables" button in the blocks palette, and then select the "Make a Variable" button. Uncheck the box by the "Speed" variable in the "Variables" section so it doesn't appear on the stage.

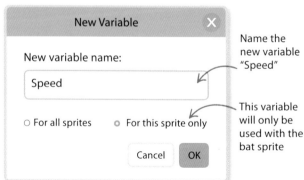

New Variable

New variable name:

Speed

○ For all sprites   ● For this sprite only

Cancel   OK

Name the new variable "Speed"

This variable will only be used with the bat sprite

7 Add the below code to the bat. In the main "forever" loop, the bat moves to a random position on the left of the stage, chooses a random speed, then moves backwards and forwards across the stage until the bananas hit it. When the bat is hit, it drops to the ground.

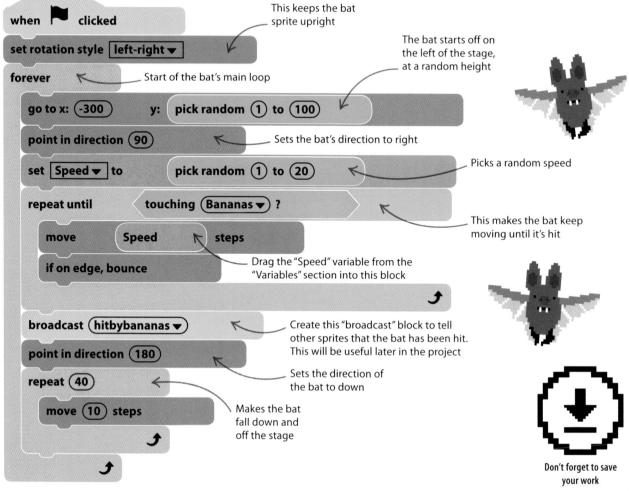

```
when [flag] clicked
set rotation style [left-right ▼]
forever
    go to x: (-300)  y: (pick random (1) to (100))
    point in direction (90)
    set [Speed ▼] to (pick random (1) to (20))
    repeat until < touching (Bananas ▼) ? >
        move (Speed) steps
        if on edge, bounce
    broadcast (hitbybananas ▼)
    point in direction (180)
    repeat (40)
        move (10) steps
```

This keeps the bat sprite upright

Start of the bat's main loop

The bat starts off on the left of the stage, at a random height

Sets the bat's direction to right

Picks a random speed

This makes the bat keep moving until it's hit

Drag the "Speed" variable from the "Variables" section into this block

Create this "broadcast" block to tell other sprites that the bat has been hit. This will be useful later in the project

Sets the direction of the bat to down

Makes the bat fall down and off the stage

Don't forget to save your work

● MONKEY MAYHEM

The finishing touches

To make the game even more exciting, you can add a timer, use a variable to keep score of how many bats the player hits, and add a game-over screen that appears once the player is out of time.

8 ▸ Create a new variable called "Time". Make sure it's available for all sprites in the game by selecting the "For all sprites" option. Check that the box next to the variable in the blocks palette is ticked, so that players can see the time displayed on the stage.

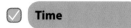

9 ▸ Click on the small picture of the stage in the stage list, then select the "Backdrops" tab above the blocks palette. Right-click the existing backdrop and duplicate it. Add the words "GAME OVER" to the new backdrop.

Use the text tool to write on the duplicate backdrop

Your game-over screen will look something like this

10 ▸ Click the "Code" tab and add this code to the stage to set up the timer. When the timer begins, it starts a count-down loop. When the loop finishes, the "GAME OVER" screen is shown and the game ends.

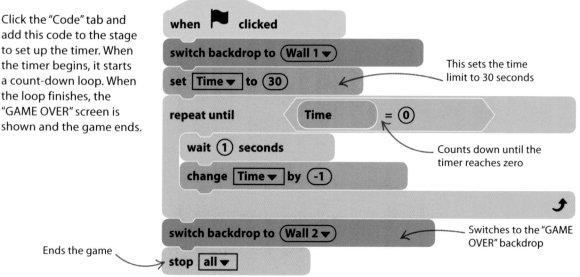

This sets the time limit to 30 seconds

Counts down until the timer reaches zero

Switches to the "GAME OVER" backdrop

Ends the game

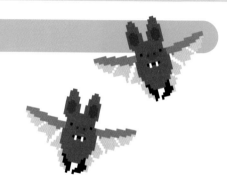

11 Click the bananas sprite in the sprites list. Create a new variable called "Score" and make it available for all sprites. Move the score to the top right of the stage by dragging it.

Tick the box to show the score on the stage

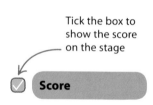

12 Add this short bit of code to the bananas sprite. It sets the score to 0 at the beginning of the game.

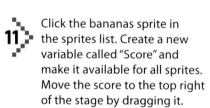

This resets the score

13 Add this code to the bananas sprite too. When the bananas hit the bat, it plays a sound, increases the score by 10, and hides the bananas.

Makes the bananas disappear

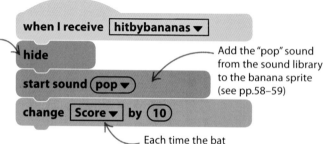

Add the "pop" sound from the sound library to the banana sprite (see pp.58–59)

Each time the bat is hit the player scores 10 points

14 Next add some music to the game. Click on the stage and select the "Sounds" tab above the blocks palette. Load the "Eggs" music from the sound library.

Add the "Eggs" music from the "Sounds" tab

15 Add the code below to the stage. It plays the "Eggs" music on a loop, but will stop when the "stop all" block ends the game.

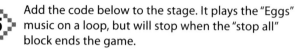
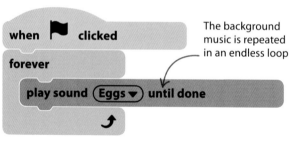

The background music is repeated in an endless loop

Don't forget to save your work

REMEMBER

Achievements

Congratulations – you've built a complete Scratch game. Here are some of the things you have achieved so far:

Made a sprite throw objects at another sprite.

Made a sprite fall off the stage once hit.

Added a time limit to your game.

Added background music that plays as long as the game continues.

Added a game-over screen that appears at the end of the game.

⦿ MONKEY MAYHEM

Time to play

Now the game is ready to play. Click the green flag to start and see how many times you can hit the bat with the bananas before the time runs out.

To make the game last longer, try increasing the time limit

You can edit the program to give the player more points for each successful hit

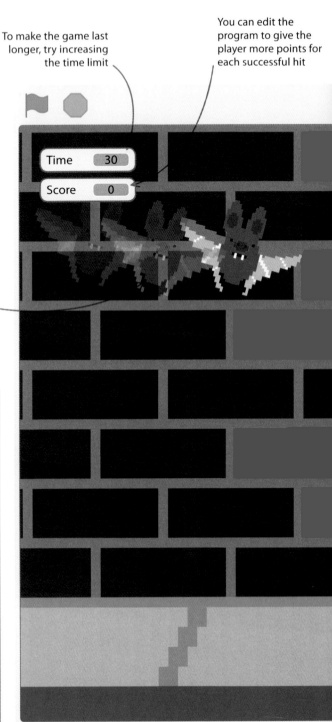

Left cursor key

Right cursor key

Spacebar

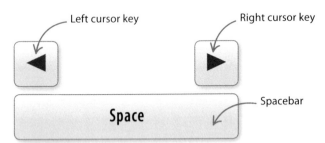

△ **Controls**
Steer the monkey left and right with the keyboard cursor keys. Tap the spacebar to fire bananas at the bat.

To make the game harder, make the bat move faster

⠿ **EXPERT TIPS**

Adding more sprites

To add more bats to aim at, right-click the bat in the sprites list and select "duplicate". A new bat will appear with the same code as the first one. Try adding some other flying sprites:

1. Add a sprite from the sprites library. The flying hippo ("Hippo1") is great for this game.

2. Click on the bat in the sprites list.

3. Click the bat's code and hold the mouse button down.

4. Drag the bat's code on to the new sprite in the sprites list.

5. The code will copy across to the new sprite.

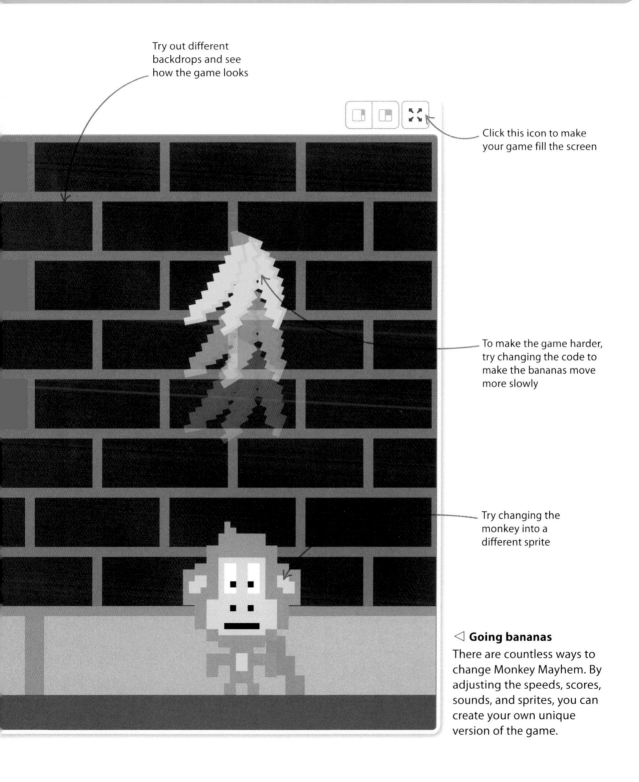

Try out different
backdrops and see
how the game looks

Click this icon to make
your game fill the screen

To make the game harder,
try changing the code to
make the bananas move
more slowly

Try changing the
monkey into a
different sprite

◁ **Going bananas**
There are countless ways to
change Monkey Mayhem. By
adjusting the speeds, scores,
sounds, and sprites, you can
create your own unique
version of the game.

Time to experiment

Now you've learned the basics of Scratch, you can experiment with some of its more advanced features. The more you practise, the better your coding will become.

SEE ALSO

What is **86–87 〉**
Python?

Simple **102–103 〉**
commands

Things to try

Not sure what to do next with Scratch? Here are a few ideas. If you don't feel ready to write a whole program on your own yet, you can start with one that has already been written and change parts of it.

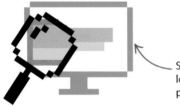

Scratch allows you to look at the coding of all projects on its website

△ **Join a coding club**
Is there a coding club in your school or local area? They're great places to meet other Scratch users and share ideas.

△ **Look at code**
Looking at other programs is a great way to learn. Go through projects shared on the Scratch website. What can you learn from them?

▷ **Remix existing projects**
Can you improve the projects on the Scratch website? Scratch lets you add new features and then share your version.

Backpack

The backpack enables you to store useful code, sprites, sounds, and costumes and move them from project to project. It's found at the bottom of the Scratch screen.

Drag and drop to copy code or sprites into the backpack

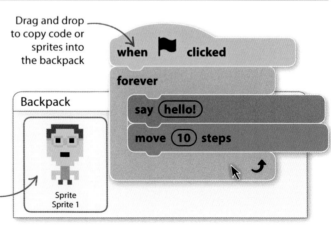

```
when ⚑ clicked
forever
    say (hello!)
    move (10) steps
```

Backpack

A sprite in the backpack

Sprite
Sprite 1

▷ **Drag and drop**
You can drag sprites and code into your backpack, then add them to other projects.

Tutorials

Scratch has built-in tutorials for learning more of the basics. A tutorial is a step-by-step lesson that teaches you how to do something.

1 **Choose a tutorial**
Click the tutorials icon. Look through the list of projects that appear to see what you'd like to work on. There are lots to choose from.

Select this icon to open the tutorial library

2 **Start learning**
Click on a tutorial to get started. Scratch will tell you what to do at each step.

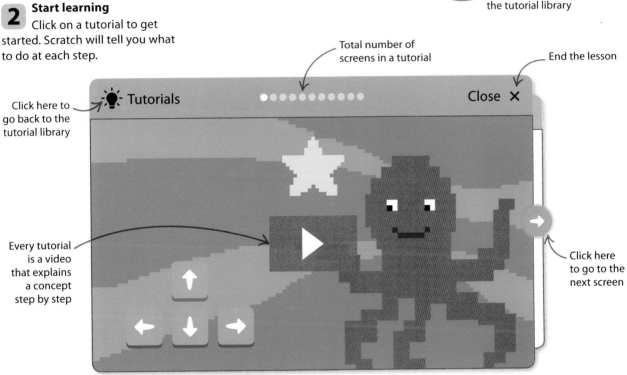

Click here to go back to the tutorial library

Total number of screens in a tutorial

End the lesson

Every tutorial is a video that explains a concept step by step

Click here to go to the next screen

Learn another language

You're now on your way to mastering your first programming language. Learning other languages will enable you to write different types of programs. Why not try Python next? What you've already learned about Scratch will help you to pick up Python quickly.

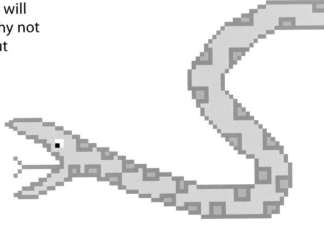

▷ **Similar to Scratch**
Python uses loops, variables, and branches too. Use your Scratch knowledge to start learning Python!

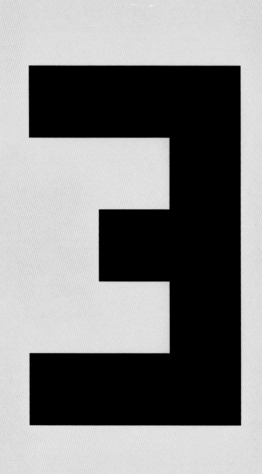

Playing
with Python

What is Python?

Python is a text-based programming language. It takes a bit longer to learn than Scratch, but can be used to do much more.

SEE ALSO

Installing **88–91)**
Python

Simple **102–103)**
commands

Harder **104–105)**
commands

A useful language

Python is a versatile language that can be used to make many different types of programs, from word processing to web browsers. Here are a few great reasons to learn Python.

1 **Easy to learn and use**
Python programs are written in a simple language. The code is quite easy to read and write, compared to many other programming languages.

2 **Contains ready-to-use code**
Python contains libraries of preprogrammed code that you can use in your programs. It makes it easier to write complex programs quickly.

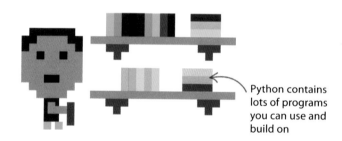

Python contains lots of programs you can use and build on

3 **Useful for big organizations**
Python is powerful. It can be used to write real-world programs. It is used by Google, NASA, and Pixar, among others.

• • EXPERT TIPS

Getting started

Before learning how to program in Python, it's useful to get familiar with how it works. The next few pages will teach you how to:

Install Python: Python is free, but you'll have to install it yourself (see pp.88–91).

Use the interface: Make a simple program and save it on the computer.

Experiment: Try some simple programs to see how they work.

Scratch and Python

Lots of elements that are used in Scratch are also used in Python – they just look different. Here are a few similarities between the two languages.

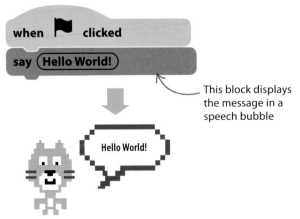

This block displays the message in a speech bubble

△ **Print in Scratch**
In Scratch, the "say" block is used to show something on the screen.

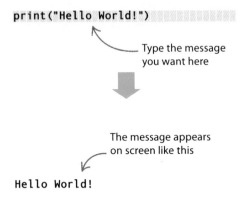

```
print("Hello World!")
```

Type the message you want here

The message appears on screen like this

```
Hello World!
```

△ **Print in Python**
In Python, a command called "print" displays text on the screen.

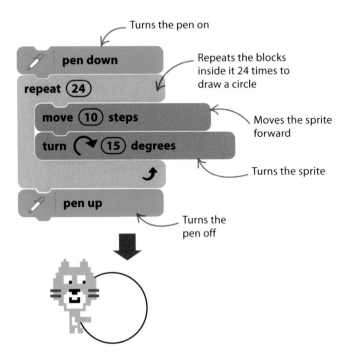

Turns the pen on

Repeats the blocks inside it 24 times to draw a circle

Moves the sprite forward

Turns the sprite

Turns the pen off

△ **Turtle graphics in Scratch**
The code above uses the "pen down" block to move the cat sprite and draw a circle.

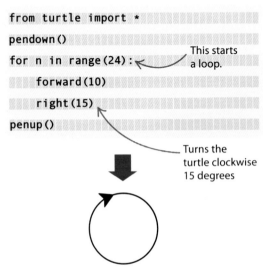

```
from turtle import *
pendown()
for n in range(24):
    forward(10)
    right(15)
penup()
```

This starts a loop.

Turns the turtle clockwise 15 degrees

△ **Turtle graphics in Python**
There's also a turtle in Python. The code above can be used to draw a circle.

Installing Python

Before you can use the Python programming language, you need to download and install it on your computer. Python 3 is free, easy to install, and works on Windows PCs, Macs, and Linux operating systems such as Ubuntu.

What is IDLE?

When you install Python 3, you'll also get a free program called IDLE (Integrated DeveLopment Environment). Designed for beginners, IDLE includes a basic text editor that allows you to write and edit Python code.

⊙ ⊙ EXPERT TIPS

Saving code

When saving work in Python, you will need to use the "File > Save As..." menu command so you can name your files. First create a folder to keep all your files in. Give the folder a clear name, like "PythonCode", and agree with the person who owns the computer where to keep it.

WINDOWS

△ **Windows**

Before you download Python, check what kind of operating system your computer has. If you have Windows, find out whether it's the 32-bit or 64-bit version. Click the "Start" button, right-click "This PC", and then left-click "Properties".

MAC

△ **Mac**

If you use an Apple Mac, find out which operating system it has before you install Python. Click the apple icon in the top left and choose "About This Mac".

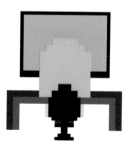

UBUNTU

△ **Ubuntu**

Ubuntu is a free operating system that works just like Windows and Macs. To find out how to install Python on Ubuntu, turn over to page 91.

Python 3 on Windows

Before you install Python 3 on a Windows PC, make sure you get permission from the computer's owner. You may also need to ask the owner to provide an admin password during installation.

1 Go to the Python website
Type the address below into your internet browser to open the Python website. Click on "Downloads" to open the download page.

Q http://www.python.org

This is the URL (web address) for Python

2 Download Python
Click on the latest version of Python for Windows, beginning with the number 3. You can choose either the web-based installer or the executable installer.

Choose this if you have a 32-bit version of Windows

Don't worry about the exact number, as long as it has a 3 at the front

- Python 3.7.0 - 2019-02-15
  - Download Windows x86 web-based installer
  - Download Windows x86-64 web-based installer

Choose this if you have a 64-bit version of Windows

3 Install
The installer file will download automatically. When it finishes, double-click it to install Python. Choose "install for all users" and click "next" at each prompt, without changing the default settings.

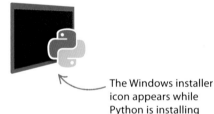

The Windows installer icon appears while Python is installing

4 Run IDLE
Now check that the program installed correctly. Open the Windows "Start" menu, choose "All Apps", then select "IDLE".

Click on this icon to run the IDLE application

5 A Python window opens
A window like the one below should open up. You can now start coding – just type into the window after the angle brackets (>>>).

Python 3.7.0 Shell
IDLE   File   Edit   Shell   Debug   Window   Help
Python 3.7.0 (v3.7.0:5fd3365926, Feb 15 2019, 13:38:16) [MSC v.1900 32 bit (Intel)] on win32
Type "copyright", "credits" or "license()" for more information.
>>>   Begin typing code here

Python 3 on a Mac

Before you install Python 3 on a Mac, make sure you get permission from the computer's owner. You may also need to ask the owner to provide an admin password during installation.

1 **Go to the Python link**
Type the address below into your web browser to open the Python website. Click on "Downloads" in the navigation panel to go to the download page.

Q http://www.python.org

Don't worry about the exact number, as long as it has a 3 at the front

2 **Download Python**
Check which operating system your Mac has (see page 88) and click on the matching version of Python 3. You'll be prompted to save a .dmg file. Save it on your Mac desktop.

- Python 3.7.0 - 2019-02-15
  - Download macOS 64-bit/32-bit installer

This version runs on most Macs

3 **Install**
Double-click the .dmg file. A window will open with several files in it, including the Python installer file "Python.pkg". Double-click it to start the installation.

Python installer file

Python.pkg

4 **Run IDLE**
During installation, click "next" at each prompt to accept the default settings. After installation ends, open the "Applications" folder on your Mac and open the "Python" folder (make sure you select Python 3, not Python 2). Double-click "IDLE" to check the installation worked.

IDLE icon

5 **A Python window opens**
A window like the one below should open. You can now start coding – just type into the window after the angle brackets.

```
                        Python 3.7.0 Shell
IDLE    File    Edit    Shell    Debug    Window    Help
Python 3.7.0 (v3.7.0:1bf9cc5093, Feb 15 2019, 13:38:16)
[Clang 6.0 (clang-600.057)] on darwin
Type "copyright", "credits" or "license()" for more information.
>>>
```

Python 3 on Ubuntu

Most recent versions of Ubuntu come with Python and IDLE pre-installed. If you can't find IDLE on your machine, you can download it without having to use a browser – just follow the steps below. If you have a different version of Linux, ask the computer's owner to install IDLE for you.

1 **Go to Ubuntu Software Centre**
Find the Ubuntu Software Centre icon in the Dock or the Dash and double-click it.

2 **Enter "Terminal" into the search bar**
You will see a search icon in the top right. Type "Terminal" in the box and press enter to open Ubuntu's command line.

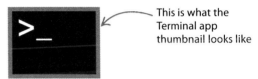

This is what the Terminal app thumbnail looks like

3 **Install IDLE**
Type the commands shown below into the Terminal window one at a time. Make sure you start typing after the "$" sign.

This updates the system so you can install the latest version of any software

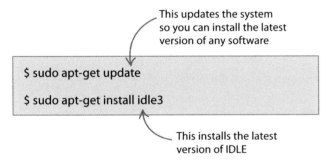

```
$ sudo apt-get update

$ sudo apt-get install idle3
```

This installs the latest version of IDLE

4 **Run IDLE**
After the installation is complete, enter "IDLE" into the search bar and double-click on the blue-and-yellow "IDLE" icon.

IDLE icon

5 **A Python window opens**
A window like the one below should open. You can now start coding – just type into the window after the angle brackets.

```
                        Python 3.7.0 Shell
IDLE    File    Edit    Shell    Debug    Window    Help

Python 3.7.0 (default, Feb 15 2019, 18:25:56)

[Open Watcom] on linux

Type "help", "copyright", "credits" or "license()" for more information.

>>>
```

Introducing IDLE

IDLE helps you write and run programs in Python. See how it works by creating this simple program that writes a message on the screen.

SEE ALSO

❰ **88–91** Installing Python

Which **106–107** ❱ window?

Working in IDLE

Follow these steps to make a Python program using IDLE. It will teach you how to enter, save, and run programs.

1 Start IDLE
Start up IDLE using the instructions for your computer's operating system (see pp.88–91). The shell window opens. This window shows the program output (any information the program produces) and any errors.

EXPERT TIPS

Different windows

Python uses two different windows – the "shell" window and the "code" window (see pages 106–107). We've given them different colours to tell them apart.

Shell window

Code window

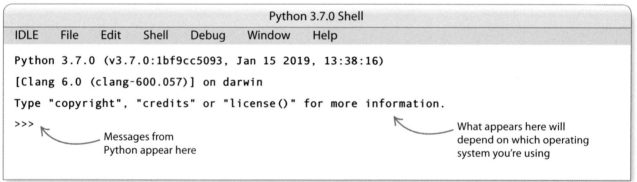

```
                    Python 3.7.0 Shell

IDLE    File    Edit    Shell    Debug    Window    Help

Python 3.7.0 (v3.7.0:1bf9cc5093, Jan 15 2019, 13:38:16)

[Clang 6.0 (clang-600.057)] on darwin

Type "copyright", "credits" or "license()" for more information.

>>>
```

Messages from Python appear here

What appears here will depend on which operating system you're using

2 Open a new window
Click the "File" menu at the top of the shell window and select "New File". This opens the code window.

This is the shell window

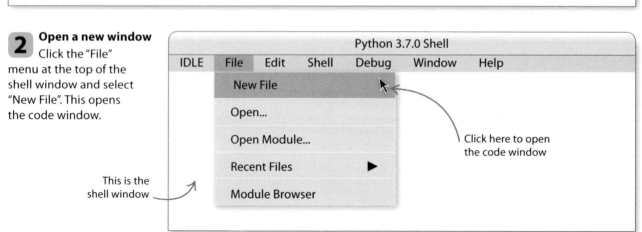

```
                    Python 3.7.0 Shell

IDLE    File    Edit    Shell    Debug    Window    Help

        New File

        Open...

        Open Module...

        Recent Files        ▶

        Module Browser
```

Click here to open the code window

3 Enter the code
In the new code window, type in this text. It's an instruction to write the words "Hello World!"

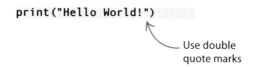

```
print("Hello World!")
```

Use double quote marks

4 Save the code window
Click the "File" menu and select "Save As...". Enter the file name "HelloWorld" and click "Save".

If you get an error message, check your code carefully to make sure you haven't made any mistakes.

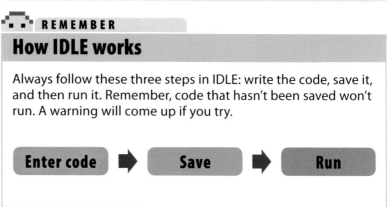

Untitled						
IDLE	File	Edit	Format	Run	Window	Help

prin   New File

Open...

Open Module...

Recent Files ▶

Module Browser

Path Browser

Close

Save

Save As...

Click here to save the file

5 Run the program
In the code window, click the "Run" menu and select "Run Module". This will run the program in the shell window.

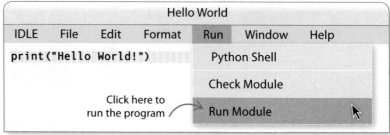

Hello World						
IDLE	File	Edit	Format	Run	Window	Help

```
print("Hello World!")
```

Python Shell

Check Module

Click here to run the program

Run Module

6 Output in the shell window
Look at the shell window. The "Hello World!" message should appear when the program runs. You've now created your first bit of code in Python!

```
>>>
Hello World!
>>>
```

The message will appear without quote marks

▪▪▪ REMEMBER
How IDLE works

Always follow these three steps in IDLE: write the code, save it, and then run it. Remember, code that hasn't been saved won't run. A warning will come up if you try.

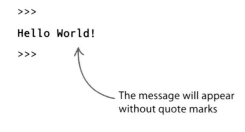

Enter code ➡ **Save** ➡ **Run**

Errors

Sometimes programs don't work the first time, but they can always be fixed. When code for a program isn't entered correctly, Python will display an error message telling you what has gone wrong.

SEE ALSO

Bugs and **148–149 ›**
debugging

What next? **176–177 ›**

Errors in the code window

When trying to run a program in the code window, you might see a pop-up window appear with an error message (such as "SyntaxError") in it. These errors stop the program from running and need to be fixed.

1 **Syntax error**
If a pop-up window appears with a "SyntaxError" message, it often means there's a spelling mistake or typing error in the code.

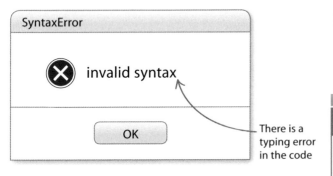

There is a typing error in the code

There is incorrect spacing in the code, which is preventing the program from running

2 **Error highlighted**
Click "OK" in the pop-up window and you'll go back to your program. There will be a red highlight on or near the error. Check that line for mistakes carefully.

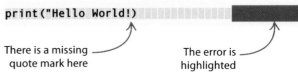

```
print("Hello World!)
```

There is a missing quote mark here

The error is highlighted

• • (EXPERT TIPS

Classic errors

Some mistakes are particularly easy to make. Keep an eye out for these common problems:

Upper vs lower case: The case has to match exactly. If you write "Print" instead of "print", Python won't understand the instruction.

Single and double quotes: Don't mix up single and double quotes. All opening quotes need a matching closing quote.

Minus and underscore: Don't confuse the minus sign (-) with the underscore sign (_).

Different brackets: Different-shaped brackets, such as (), {} and [], are used for different things. Use the correct ones, and check there's a complete pair.

Errors in the shell window

Sometimes, an error message will appear in red text in the shell window. This will also stop the program from working.

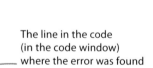
Red text means there's something wrong!

1 Name error
If the error message "NameError" appears, it means Python can't understand one of the words that has been used. If the error is in code entered in the code window, right-click on the error message in the shell window and select "Go to file/line".

```
>>>
Traceback (most recent call last):
    File "C:\PythonCode\errors.py", line 1, in <module>
        pront("Hello World!")
NameError: name "pront" is not defined
```

The line in the code (in the code window) where the error was found

The word Python doesn't understand

```
Cut
Copy
Paste
Go to file/line
```

Click here to highlight the line where the error appears in the code window

2 Fix the error
The line with the error is highlighted in the code window. The word "pront" has been typed instead of "print". You can then edit the code to fix the error.

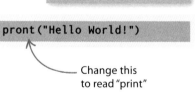
```
pront("Hello World!")
```

Change this to read "print"

Spotting errors

Use the tips on these two pages to find the line in the code where the errors appear, then double-check that line. Go through the check list on the right to help you find out what has gone wrong.

▷ **When things go wrong**
There are some methods you can use to find errors more easily. Here's a handy check list.

ERROR BUSTING	
Check your code for the following points	✓
Have you copied exactly what you were asked to enter?	✓
Have you spelled everything correctly?	✓
Are there two quote marks ("") around the expression you want to print?	✓
Do you have extra spaces at the beginning of the line? Spacing is very important in Python.	✓
Have you checked the lines above and below the highlighted line? Sometimes that's where the problem is.	✓
Have you asked someone else to check the code against the book? They might spot something you have missed.	✓
Are you using Python 3 not Python 2? Programs for Python 3 don't always work in Python 2.	✓

SEE ALSO

Ghost game **98–99 ⟩**
decoded

Program **100–101 ⟩**
flow

⟩ PROJECT 4

Ghost game

This simple game highlights some of the things to watch out for when writing programs in Python. Once the code has been typed in, run the program to play the game. Can you escape the haunted house?

Use double quotes

1 Start IDLE, and use the "File" menu to open a new window. Save the game as "ghostgame". Arrange the windows so you can see them both, then type this into the code window.

These must be underscores, not minus signs

This section needs to be indented by four spaces. If this doesn't happen automatically, check there is a colon after "feeling_brave"

This indent will start at eight spaces and needs to be reduced to just four spaces

Delete all indents here

```python
# Ghost Game
from random import randint
print("Ghost Game")
feeling_brave = True
score = 0
while feeling_brave:
    ghost_door = randint(1, 3)
    print("Three doors ahead...")
    print("A ghost behind one.")
    print("Which door do you open?")
    door = input("1, 2, or 3?")
    door_num = int(door)
    if door_num == ghost_door:
        print("GHOST!")
        feeling_brave = False
    else:
        print("No ghost!")
        print("You enter the next room.")
        score = score + 1
print("Run away!")
print("Game over! You scored", score)
```

Only use capital letters where they are shown

Make sure to add a colon here

Use two equals signs here

There should be no quotes around "score" here

2 Once the code has been carefully typed in, use the "Run" menu to select "Run Module". You must save the program first.

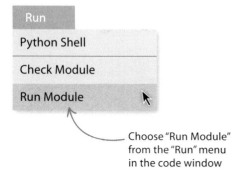

Choose "Run Module" from the "Run" menu in the code window

 3 The game begins in the shell window. The ghost is hiding behind one of three doors. Which one will you pick? Type 1, 2, or 3 then press "Enter".

```
Ghost Game
Three doors ahead...
A ghost behind one.
Which door do you open?
1, 2, or 3?
```

Type in your guess

4 The aim of the game is to pick a door with no ghost behind it. If this happens, you'll move to the next room and keep playing the game.

```
Ghost Game
Three doors ahead...
A ghost behind one.
Which door do you open?
1, 2, or 3?3
No ghost!
```

The number you type in appears here

This is what you'll see if there is no ghost behind the door you choose

5 If you're unlucky you'll pick a door with a ghost behind it, and the game ends. Run the program again to see if you can beat your last score.

```
Ghost Game
Three doors ahead...
A ghost behind one.
Which door do you open?
1, 2, or 3?2
GHOST!
Run away!
Game over! You scored 0
```

This is what appears if the ghost is behind your door

The score shows how many rooms you survived

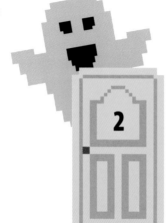

Ghost game decoded

The ghost game displays some of the key features of Python. You can break down the code to see how the program is structured and what the different parts of it do.

SEE ALSO

❮ **96–97** Ghost game

Program **100–101** ❯
flow

Code structure

Python uses spaces at the start of lines to work out which instructions belong together. These spaces are called "indents". For example, the code after "while feeling_brave" is indented by four spaces to show it's all part of the main loop.

```
# Ghost Game
from random import randint          1
print("Ghost Game")
feeling_brave = True
score = 0
while feeling_brave:
    ghost_door = randint(1, 3)
    print("Three doors ahead...")
    print("A ghost behind one.")    2
    print("Which door do you open?")
    door = input("1, 2 or 3?")
    door_num = int(door)
    if door_num == ghost_door:
        print("GHOST!")
        feeling_brave = False
    else:                           3
        print("No ghost!")
        print("You enter the next room.")
        score = score + 1
print("Run away!")                  4
print("Game over! You scored", score)
```

Game set-up

The main loop

◁ **Code key**

This diagram shows the structure of the ghost game. The numbered parts are explained in more detail below.

Branching part

Game ending

This is a "comment". It's not shown when the game is run

1 **Game set-up**
These instructions only run once – at the beginning of the game. They set up the title, variables, and the "randint" command.

```
# Ghost Game
from random import randint
print("Ghost Game")
feeling_brave = True
score = 0
```

This sets up the "randint" command, which generates random numbers

The "print" command displays text when the game is run

This resets the score to 0

● ● ❱ **EXPERT TIPS**

Type carefully

When using Python, enter the code very carefully. If you miss out a colon, quote mark, or bracket, the program won't work properly. You need to match the use of capital letters and spaces exactly too.

2 **The main loop**
This loop tells the story and receives the player's guess. It keeps on going as long as there isn't a ghost behind the door that's picked. When a ghost appears, the "feeling_brave" variable changes to "False" and the loop stops repeating.

3 **Branching part**
The program takes a different path depending on whether or not there was a ghost behind the door that was picked. If there was a ghost, the "feeling_brave" variable is set to "False". If there wasn't a ghost, the player's score increases by one.

```python
while feeling_brave:
    ghost_door = randint(1, 3)
    print("Three doors ahead...")
    print("A ghost behind one.")
    print("Which door do you open?")
    door = input("1, 2 or 3?")
    door_num = int(door)
    if door_num == ghost_door:
        print("GHOST!")
        feeling_brave = False
    else:
        print("No ghost!")
        print("You enter the next room.")
        score = score + 1
```

This selects a random number between 1 and 3

The "print" command displays the text onscreen

This line asks for the player's answer

This branch runs if there's a ghost behind the door the player picks

If there's no ghost, the player sees this message

The score increases by one each time the player enters a room without meeting a ghost

This shows a message telling the player to run away from the ghost

4 **Game ending**
This runs just once, when you meet the ghost and the loop ends. Python knows this isn't part of the loop because it's not indented.

```python
print("Run away!")
print("Game over! You scored", score)
```

The score is a variable – it will change depending on how many rooms the player gets through

Achievements

Congratulations – you've created your first Python game! You'll learn more about these commands later in the book, but you've already achieved a lot:

Entered a program: You've typed a program into Python and saved it.

Run a program: You've learned how to run a Python program.

Structured a program: You've used indents to structure a program.

Used variables: You've used variables to store the score.

Displayed text: You've displayed messages on the screen.

Program flow

Before learning more about Python, it's important to understand how programs work. The programming basics learned in Scratch can also be applied to Python.

SEE ALSO

❮ **30–31** Coloured blocks and code

Simple **102–103** ❯ commands

Harder **104–105** ❯ commands

From input to output

A program takes input (information in), processes it (or changes it), and then gives back the results (output). It's a bit like a chef taking ingredients, turning them into cakes, and then giving you the cakes to eat.

Input

Processing

Output

Input command

Keyboard

Mouse

Variables

Maths

Loops

Branches

Functions

Print command

Screen

Graphics

△ **Program flow in Python**
In Python, the keyboard and mouse are used to input information, which is processed using elements such as loops, branches, and variables. The output is then displayed on the screen.

Looking at the Ghost game through Scratch goggles

Program flow works the same in most programming languages. Here are some examples of input, processing, and output in Python's Ghost game – and what they might look like in Scratch.

Python and Scratch are more similar than they appear.

1 Input

In Python, the "input()" function takes an input from the keyboard. It's similar to the "ask and wait" block in Scratch.

```
door = input("1, 2 or 3?")
```

The question appears on screen

The question in the Scratch block

```
ask (1, 2 or 3?) and wait
```

"ask and wait" Scratch block

2 Processing

Variables are used to keep track of the score and the function "randint" picks a random door. Different blocks are used to do these things in Scratch.

```
score = 0
```

Sets the variable "score" to 0

This Scratch block sets the value of the variable "score" to 0

```
set [score ▼] to [0]
```

"set score to 0" Scratch block

```
ghost_door = randint(1, 3)
```

Selects a random whole number between 1 and 3

```
pick random (1) to (3)
```

"pick random" Scratch block

This Scratch block selects a random number

3 Output

The "print()" function is used to output things in Python, while the "say" block does the same thing in Scratch.

Displays "Ghost game" on the screen

```
print("Ghost game")
```

Shows a speech bubble containing the words "Ghost game"

```
say (Ghost game)
```

"say" Scratch block

Simple commands

At first glance, Python can look quite scary, especially when compared to Scratch. However, the two languages aren't actually as different as they seem. Here is a guide to the similarities between basic commands in Python and Scratch.

SEE ALSO

❰ **86–87** What is Python?

Harder **104–105** ❱ commands

Command	Python 3	Scratch 3.0
Run program	"Run" menu or press "F5" (in code window)	⚑
Stop program	Press "CTRL-C" (in shell window)	⬛
Write text to screen	`print("Hello!")`	say (Hello!)
Set a variable to a number	`magic_number = 42`	set [magic_number ▾] to (42)
Set a variable to a text string	`word = "dragon"`	set [word ▾] to (dragon)
Read text from keyboard into variable	`age = input("age?")` `print("I am " + age)`	ask (age?) and wait say join (I am) answer

Add a number to a variable	`cats = cats + 1`
	or
	`cats += 1`
		change [cats ▾] by (1)

Add	`a + 2`	a + (2)
Subtract	`a - 2`	a − (2)
Multiply	`a * 2`	a * (2)
Divide	`a / 2`	a / (2)

Command	Python 3	Scratch 3.0

Forever loop	`while True:`
	    `jump()`
		forever
		  jump

Loop 10 times	`for i in range (10):`
	    `jump()`
		repeat (10)
		  jump

Is equal to?	`a == 2`	a = (2)
Is less than?	`a < 2`	a < (2)
Is more than?	`a > 2`	a > (2)
NOT	`not`	not
OR	`or`	or
AND	`and`	and

If then	`if a == 2:`
	    `print("Hello!")`
		if a = (2) then
		  say Hello!

If then else	`if a == 2:`
	    `print("Hello!")`
	`else:`
	    `print("Goodbye!")`
		if a = 2 then
		  say Hello!
		else
		  say Goodbye!

Harder commands

Python can also be used to do some of the more complicated things that are possible in Scratch: for example, creating complex loops, playing with strings and lists, and drawing pictures with turtle graphics.

SEE ALSO

‹ **86–87** What is Python?

‹ **102–103** Simple commands

Command	Python 3	Scratch 3.0
Loops with conditions	```while roll != 6: jump()```	repeat until (roll = 6) jump
Wait	```from time import sleep sleep(2)```	wait (2) seconds
Random numbers	```from random import randint roll = randint(1, 6)```	set [roll ▼] to (pick random (1) to (6))
Define a function or subprogram	```def jump(): print("Jump!")```	define jump / think (Jump!)
Call a function or subprogram	```jump()```	jump
Define a function or subprogram with input	```def greet(who): print("Hello " + who)```	define greet (who) / say (join (Hello) (who))
Call a function or subprogram	```greet("chicken")```	greet (chicken)

Command	Python 3	Scratch 3.0

Turtle graphics	```from turtle import *```
	```clear()```
	```pendown()```
	```forward(100)```
	```right(90)```
	```penup()```
		erase all
		pen down
		move (100) steps
		turn ↻ (90) degrees
		pen up

Join strings	```print(greeting + name)```	say join (greeting) (name)
Get one letter of a string	```name[0]```	letter (1) of (name)
Length of a string	```len(name)```	length of (name)
Create an empty list	```menu = list()```	Make a List
Add an item to end of list	```menu.append(thing)```	add (thing) to [menu ▼]
How many items on list?	```len(menu)```	length of [menu ▼]
Value of 5th item on list	```menu[4]```	say item (5 ▼) of [menu ▼]
Delete 2nd item on list	```del menu[1]```	delete (2 ▼) of [menu ▼]
Is item on list?	```if "olives" in menu:``` ``` print("Oh no!")```	if [menu ▼] contains (olives) then say (Oh no!)

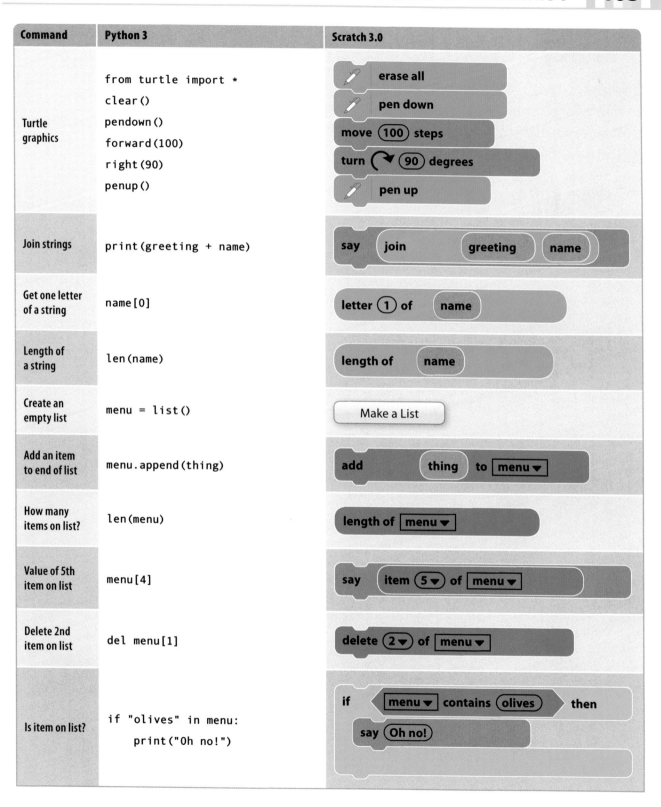

Which window?

There are two different windows to choose from in IDLE. The code window can be used to write and save programs, while the shell window runs Python instructions straight away.

SEE ALSO

‹ 92–93 Introducing IDLE

‹ 96–97 Ghost game

The code window

So far in this book, the code window has been used to write programs. You enter the program, save it, run it, and the output appears in the shell window.

▽ **Running programs**
This process is used for running Python programs. Programs always have to be saved before running them.

| Enter code | ➡ | Save | ➡ | Run module | ➡ | Output |

1 **Enter a program in the code window**
Enter this code in the code window, save it, and then click on "Run module" in the "Run" menu to run the program.

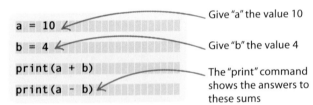

```
a = 10
b = 4
print(a + b)
print(a - b)
```

Give "a" the value 10

Give "b" the value 4

The "print" command shows the answers to these sums

2 **Output in the shell window**
When the program runs, its output (the results of the program) is shown in the shell window.

```
>>>
14
6
```

The answers to the sums appear in the shell window

The shell window

Python can also understand commands that are typed in the shell window. They run as soon as they are typed in, and the result is shown straight away.

```
>>> a = 10
>>> b = 4
>>> a + b
14
>>> a - b
6
```

Output appears immediately.

The first two commands have no output because they are just assigning values to "a" and "b"

◁ **Code and output together**
The shell window shows the code and the output together. It's easier to tell which answer belongs to which sum when the commands are typed in the shell window.

△ **Test your ideas**
The shell window gives you an immediate response, which makes it ideal for testing instructions and exploring what they can do.

Python playground

The shell window can be used to try out all sorts of Python commands, including drawing. The turtle is used to draw on screen in the same way that the pen is used in Scratch.

Loads all the commands that control the turtle

```
>>> from turtle import *
>>> forward(100)
>>> right(120)
>>> forward(100)
```

Moves the turtle forward

◁ **Enter the code**
Type these instructions in the shell window. They run after each one is typed. As the turtle moves, it draws a line.

◁ **Turtle graphic**
Can you work out how to draw other shapes, such as a square or a pentagon? To start over, type "clear()" into the shell window.

Which window should you use?

Should you use the code window or the shell window? It depends on the type of program you're writing, and whether it has to be repeated.

Code vs Shell

▷ **Code window**
The code window is ideal for longer pieces of code because they can be saved and edited. It's easier than retyping all the instructions if you want to do the same thing again or try something similar. It needs to be saved and run each time, though.

◁ **Shell window**
The shell window is perfect for quick experiments, such as checking how a command works. It's also a handy calculator. It doesn't save the instructions though, so if you're trying something you might want to repeat, consider using the code window instead.

Variables in Python

Variables are used to remember pieces of information in a program. They are like boxes where data can be stored and labelled.

SEE ALSO

Types of data **110–111** ⟩

Maths in **112–113** ⟩
Python

Strings in **114–115** ⟩
Python

Input and **116–117** ⟩
output

Functions **130–131** ⟩

Creating a variable

When a number or string is put into a variable it's called assigning a value to the variable. You use an "=" sign to do this. Try this code in the shell window.

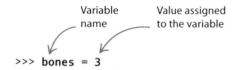

Variable name

Value assigned to the variable

```
>>> bones = 3
```

△ **Assign a number**
To assign a number, type in the variable name, an equals sign, and then the number.

Variable name

String assigned to the variable

```
>>> dogs_name = "Bruno"
```

△ **Assign a string**
To assign a string, type in the variable name, an equals sign, and then the string in quote marks.

REMEMBER

Variables in Scratch

The command to assign a variable in Python does the same job as this Scratch block. However, in Python you don't have to click a button to create a variable. Python creates the variable as soon as you assign a value to it.

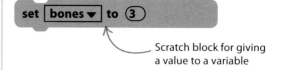

set `bones ▾` to ③

Scratch block for giving a value to a variable

Printing a variable

The "print" command is used to show something on the screen. It has nothing to do with the printer. You can use it to show the value of a variable.

```
>>> print(bones)
3
```

△ **Number output**
The variable "bones" contains the number 3, so that's what the shell window prints.

Variable name

```
>>> print(dogs_name)
Bruno
```

No quote marks here

△ **String output**
The variable "dogs_name" contains a string, so the string is printed. No quote marks are shown when you print a string.

Changing the contents of a variable

To change the value of a variable, simply assign a new value to it. Here, the variable "gifts" has the value 2. It changes to 3 when it's assigned a new value.

```
>>> gifts = 2
>>> print(gifts)
2
>>> gifts = 3
>>> print(gifts)
3
```

Changes the value of the variable

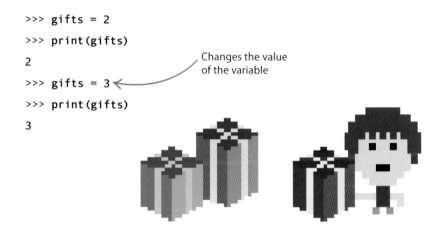

Using variables

The value of one variable can be assigned to another one using the "=" sign. For example, if the variable "rabbits" contains the number of rabbits, we can use it to assign the same value to the variable "hats", so that each rabbit has a hat.

1 **Assign the variables**
This code assigns the number 5 to the variable "rabbits". It then assigns the same value to the variable "hats".

Variable name

Value assigned to the variable

```
>>> rabbits = 5
>>> hats = rabbits
```

"hats" now has the same value as "rabbits"

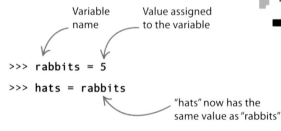

EXPERT TIPS

Naming variables

There are some rules you have to follow when naming your variables:

All letters and numbers can be used.

You can't start with a number.

Symbols such as -, /, #, or @ can't be used.

Spaces can't be used.

An underscore (_) can be used instead of a space.

Uppercase and lowercase letters are different. Python treats "Dogs" and "dogs" as two different variables.

Don't use words Python uses as a command, such as "print".

2 **Print the values**
To print two variables, put them both in brackets after the "print" command, and put a comma between them. Both "hats" and "rabbits" contain the value 5.

```
>>> print(rabbits, hats)
5 5
```

Leave a space after the comma

3 **Change the value of "rabbits"**
If you change the value of "rabbits", it doesn't affect the value of "hats". The "hats" variable only changes when you assign it a new value.

```
>>> rabbits = 10
>>> print(rabbits, hats)
10 5
```

Give "rabbits" a new value

Value for "hats" remains the same

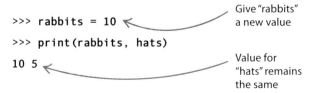

Types of data

There are several different types of data in Python. Most of the time, Python will work out what type is being used, but sometimes you'll need to change data from one type to another.

SEE ALSO

Maths in **112–113 ⟩**
Python

Strings in **114–115 ⟩**
Python

Making **118–119 ⟩**
decisions

Lists **128–129 ⟩**

Numbers

Python has two data types for numbers. "Integers" are whole numbers, (numbers without a decimal point). "Floats" are numbers with a decimal point. An integer can be used to count things such as sheep, while a float can be used to measure things such as weight.

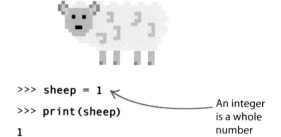

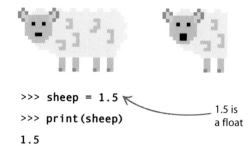

```
>>> sheep = 1
>>> print(sheep)
1
```

An integer
is a whole
number

```
>>> sheep = 1.5
>>> print(sheep)
1.5
```

1.5 is
a float

△ **Integers**
An integer is a number without a decimal point, such as the 1 in the variable "sheep".

△ **Floats**
A float is a number with a decimal point, such as 1.5. They aren't normally used to count whole objects.

Strings

Just like in Scratch, a piece of text in Python is called a "string". Strings can include letters, numbers, spaces, and symbols such as full stops and commas. They are usually put inside single quote marks.

▷ **Using a string**
To assign a string to a variable, put the text inside double quote marks.

```
>>> a = "Coding is fun!"
>>> print(a)
Coding is fun!
```

The value of the variable "a" printed out

The string in quotes

Always remember that strings need quote marks at the start and the end.

Booleans

In Python, a Boolean always has a value that is either "True" or "False". In both cases, the word begins with a capital letter.

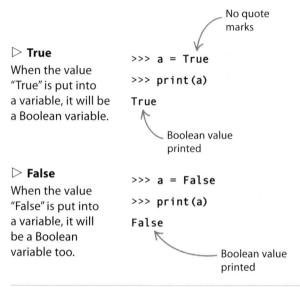

No quote marks

▷ **True**
When the value "True" is put into a variable, it will be a Boolean variable.

```
>>> a = True
>>> print(a)
True
```

Boolean value printed

▷ **False**
When the value "False" is put into a variable, it will be a Boolean variable too.

```
>>> a = False
>>> print(a)
False
```

Boolean value printed

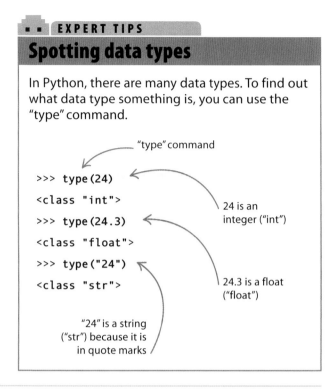

■ ■ **EXPERT TIPS**

Spotting data types

In Python, there are many data types. To find out what data type something is, you can use the "type" command.

"type" command

```
>>> type(24)
<class "int">
>>> type(24.3)
<class "float">
>>> type("24")
<class "str">
```

24 is an integer ("int")

24.3 is a float ("float")

"24" is a string ("str") because it is in quote marks

Converting data types

Variables can contain any type of data. Problems occur if you try to mix types together. Data types sometimes have to be converted, otherwise an error message will appear.

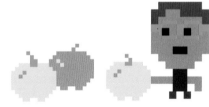

String in quote marks shown on screen

Variable name

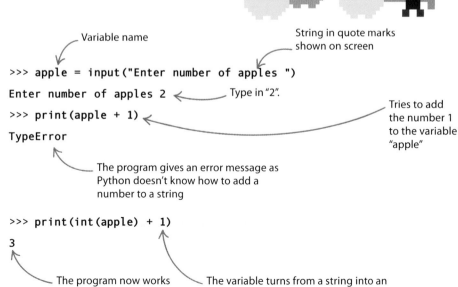

▷ **Mixed type**
The "input" command always gives a string, even if a number is entered. In this example, since "apple" actually contains a string, an error message is displayed.

```
>>> apple = input("Enter number of apples ")
Enter number of apples 2
>>> print(apple + 1)
TypeError
```

Type in "2".

Tries to add the number 1 to the variable "apple"

The program gives an error message as Python doesn't know how to add a number to a string

▷ **Converting data types**
To convert the string into a number, the "int()" command is used to turn it into an integer.

```
>>> print(int(apple) + 1)
3
```

The program now works and shows the result

The variable turns from a string into an integer, so a number can be added to it

Maths in Python

Python can be used to solve all sorts of mathematical problems, including addition, subtraction, multiplication, and division. Variables can also be used in sums.

SEE ALSO

❰ **52–53** Maths

❰ **108–109** Variables in Python

Simple calculations

In Python, simple calculations can be made by typing them into the shell window. The "print()" function is not needed for this – Python gives the answer straight away. Try these examples in the shell window:

You can't divide by zero, so you'll always get an error if you try to do so.

```
>>> 12 + 4
16
```

Use the shell window to get instant results

△ **Addition**
Use the "+" symbol to add numbers together.

The answer appears when you press "Enter"

```
>>> 12 - 4
8
```

△ **Subtraction**
Use the "-" symbol to subtract the second number from the first one.

Computers use the "*" symbol, not "x", for multiplication

```
>>> 12 * 4
48
```

△ **Multiplication**
Use the "*" symbol to multiply two numbers together.

```
>>> 12 / 4
3.0
```

Division in Python gives an answer as a float (a number with a decimal point)

△ **Division**
Use the "/" symbol to divide the first number by the second one.

Using brackets

Brackets can be used to instruct Python which part of a sum to do first. Python will always work out the value of the sum in the bracket, before solving the rest of the problem.

First it works out that 6 + 5 = 11, then 11 is multiplied by 3

```
>>> (6 + 5) * 3
33
```

△ **Addition first**
In this sum, brackets are used to instruct Python to do the addition first.

First it works out that 5 * 3 = 15, then 15 is added to 6

```
>>> 6 + (5 * 3)
21
```

Different answer

△ **Multiplication first**
Brackets here are used to do the multiplication first, in order to end up with the correct answer.

Putting answers in variables

If variables are assigned number values, you can use them within sums. When a sum is assigned to a variable, the answer goes into the variable, but not the sum.

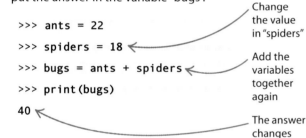

1 **Do a simple addition**
This program adds together the variables "ants" and "spiders", and puts the answer into the variable "bugs".

```
>>> ants = 22
>>> spiders = 35
>>> bugs = ants + spiders
>>> print(bugs)
57
```

Adds the values of the two variables together

Prints the value in "bugs"

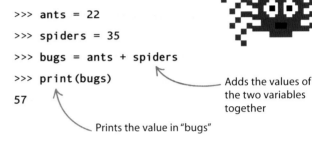

2 **Change the value of a variable**
Change the value of the "ants" or "spiders" variable. Add the variables together again and put the answer in the variable "bugs".

```
>>> ants = 22
>>> spiders = 18
>>> bugs = ants + spiders
>>> print(bugs)
40
```

Change the value in "spiders"

Add the variables together again

The answer changes

3 **Skipping the assignment**
If the sum is not assigned to the variable "bugs", even if the value of "ants" and "spiders" changes, the value of "bugs" won't.

```
>>> ants = 11
>>> spiders = 17
>>> print(bugs)
40
```

Prints the value in "bugs"

The answer hasn't changed (it's still 18 + 22)

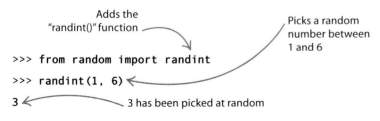

Random numbers

To pick a random number, you first need to load the "randint" function into Python. To do this, use the "import" command. The "randint()" function is already programmed with code to pick a random integer (whole number).

Adds the "randint()" function

Picks a random number between 1 and 6

```
>>> from random import randint
>>> randint(1, 6)
3
```

3 has been picked at random

△ **Roll the dice**
The "randint()" function picks a random number between the two numbers in the brackets. In this program, "randint(1, 6)" picks a value between 1 and 6.

REMEMBER

Random block

The "randint()" function works like the "pick random" block in Scratch. In Scratch, the lowest and highest possible numbers are typed into the windows in the block. In Python, the numbers are put in brackets, separated by a comma.

pick random ① to ⑥

△ **Whole numbers**
Both the Python "randint()" function and the Scratch block pick a random whole number – the result is never in decimals.

Strings in Python

Python is excellent for using words and sentences within programs. Different strings (sequences of characters) can be joined together, or individual parts of them can be selected and pulled out.

SEE ALSO

❮ **54–55** Strings and lists

❮ **110–111** Types of data

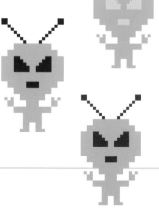

Creating a string

A string might include letters, numbers, symbols, or spaces. These are all called characters. Strings can be placed in variables.

The quote marks indicate the variable contains a string

▷ **Strings in variables**
Variables can store strings. Type these two strings into the variables "a" and "b".

```
>>> a = "Run! "
```
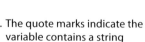
```
>>> b = "Aliens are coming."
```

Adding strings

Adding two numbers together creates a new number. In the same way, when two strings are added together, one string simply joins on to the other one.

```
>>> c = a + b
>>> print(c)
Run! Aliens are coming.
```
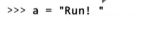

The variables "a" and "b" combine to become variable "c"

△ **Adding strings together**
The "+" symbol joins one string to another. and the answer becomes the variable "c".

A new string is added to variable "c"

```
>>> c = b + " Watch out! " + a
>>> print(c)
Aliens are coming. Watch out! Run!
```

The new string appears in the middle of the message

△ **Adding another string in between**
A new string can also be added between two strings. Try the example above.

EXPERT TIPS

Length of a string

The "len()" function is used to find out the length of a string. Python counts all of the characters, including spaces, to give the total number of characters in a string.

Calculates the length of the string in variable "a" ("Run! ")

```
>>> len(a)
4
>>> len(b)
18
```

The string in variable "b" ("Aliens are coming.") is 18 characters long

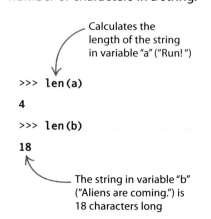

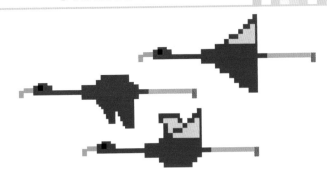

Numbering the characters

Each character in a string is allocated a number according to its position. This position number can be used to look at individual letters or symbols, or to pull them out of a string.

1 Count begins from zero
When counting the positions, Python starts at 0. The second character is in position 1, the third in position 2, and so on.

```
>>> a = "FLAMINGO"
```

The sixth letter, "N", is in position 5

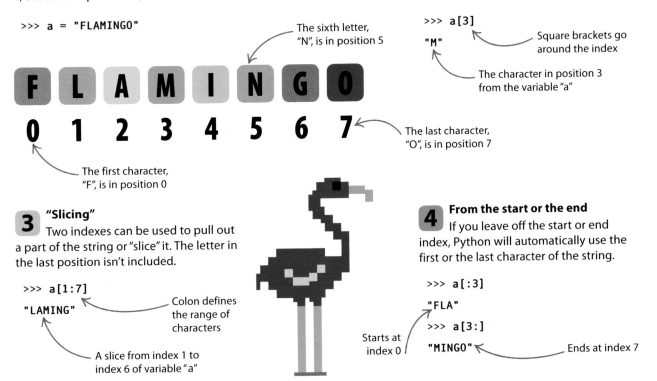

The first character, "F", is in position 0

2 Counting the characters
The position number is called an "index". It can be used to pull out a particular letter from a string.

```
>>> a[3]
"M"
```

Square brackets go around the index

The character in position 3 from the variable "a"

The last character, "O", is in position 7

3 "Slicing"
Two indexes can be used to pull out a part of the string or "slice" it. The letter in the last position isn't included.

```
>>> a[1:7]
"LAMING"
```

Colon defines the range of characters

A slice from index 1 to index 6 of variable "a"

4 From the start or the end
If you leave off the start or end index, Python will automatically use the first or the last character of the string.

```
>>> a[:3]
"FLA"
>>> a[3:]
"MINGO"
```

Starts at index 0

Ends at index 7

Apostrophes

Strings can go in single quotes or double quotes. However, the string should start and end with the same type of quote mark. This book uses double quotes. But what happens if you want to use an apostrophe in your string?

```
>>> print('It\'s a cloudy day.')
It's a cloudy day.
```

The apostrophe is included in the string

△ **Escaping the apostrophe**
So Python doesn't read an apostrophe as the end of the string, type a "\" before it. This is called "escaping" it.

Input and output

Programs interact with users through input and output. Information can be input into a program using a keyboard. Output is shown as information printed on the screen.

SEE ALSO

‹ 100–101 Program flow

‹ 110–111 Types of data

Loops **122–123 ›** in Python

Input

The "input()" function is used to accept input from the keyboard into a program. It waits until the user finishes typing and presses the "return" or "Enter" key.

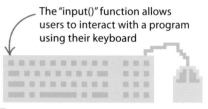

The "input()" function allows users to interact with a program using their keyboard

1 Using input
A program can prompt the user what to type. The message is put inside the brackets of "input()".

Adding a space after the colon makes the output look tidier

```
name = input ("Enter your name: ")
print("Hello", name)
```

What the program outputs depends on what name the user types

2 Output in the shell window
When the program is run, the message "Enter your name: " and its response appear in the shell window.

```
Enter your name: Jina
Hello Jina
```

Program outputs message

User types in their name

Output

The "print()" function is used to display characters in the shell window. It can be used to show a combination of text and variables.

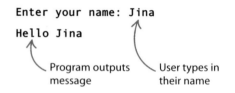

Output is displayed on the screen

1 Create some variables
Set up three variables for this simple experiment. Two are strings and one is an integer (whole number).

```
>>> a = "Dave"
>>> b = "is"
>>> c = 12
```

Quote marks show these are strings

No quote marks as this is an integer

2 Using the "print()" function
You can put several items inside the brackets of the "print()" function. You can combine variables of different types, and even combine strings and variables.

```
>>> print(a, b, c)
Dave is 12
>>> print("Goodbye", a)
Goodbye Dave
```

Comma separates the different items

Two ways to separate strings

So far, the output has been printed on one line with a space between the items. Here are two other ways of separating strings.

The separator

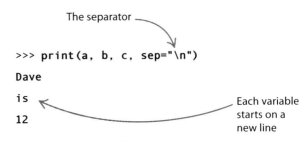

```
>>> print(a, b, c, sep="\n")
Dave
is
12
```

Each variable starts on a new line

```
>>> print(a, b, c, sep="-")
Dave-is-12
```

The character between the outputs

△ **Hyphenate the outputs**
A hyphen can be put between the variables when they're printed. Other characters, such as "+" or "*", can be used too.

△ **Outputs on new lines**
The space or character between the outputs is called a "separator" ("sep"). Using "\n" prints each output on a new line.

Three ways to end output

There are several different ways you can signal the end of the output of a "print" function.

```
>>> print(a, ".")
Dave .
```

Full stop added as a string

```
>>> print(a, end=".")
Dave.
```

Full stop added as an "end" character

△ **Add a full stop to the output**
A full stop can be added as another string to be printed, but it will print with a space before it. To avoid this, use "end="."" instead.

• • ■   EXPERT TIPS

Options at the end

The "end" and "sep" labels tell Python that the next item in the program isn't just another string. Remember to use them, otherwise the program will not work correctly.

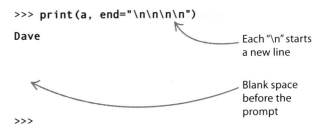

Loop to print three times

Space as "end" character

```
>>> for n in range(3):
    print("Hurray!", end=" ")
Hurray! Hurray! Hurray!
```

Output is all printed on one line

△ **Output on one line**
Usually, each new "print" command starts on a new line. To get the output all on one line use a space as the "end" character.

```
>>> print(a, end="\n\n\n\n")
Dave


>>>
```

Each "\n" starts a new line

Blank space before the prompt

△ **Blank lines at the end**
Using "\n" starts each output from a new line. Several of them can be used together to add blank lines at the end of a program.

Making decisions

Programs make decisions about what to do by comparing variables, numbers, and strings using Boolean expressions. These give an answer of either "True" or "False".

SEE ALSO

❬ 62–63 True or false?

❬ 108–109 Variables in Python

Logical operators

Logical operators are used to compare variables against numbers or strings, or even against other variables. The resulting answer is either "True" or "False".

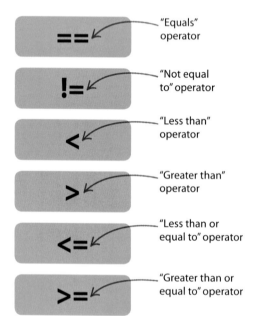

== — "Equals" operator

!= — "Not equal to" operator

< — "Less than" operator

> — "Greater than" operator

<= — "Less than or equal to" operator

>= — "Greater than or equal to" operator

△ **Types of comparison operators**
There are six comparison operators. Python uses two equals signs to compare if two things are the same. (A single equals sign is used to assign a value to a variable.)

▷ **Use the shell to check**
Logical operators also work in the shell window. Use this example to try out several logical operators, including "not", "or", and "and".

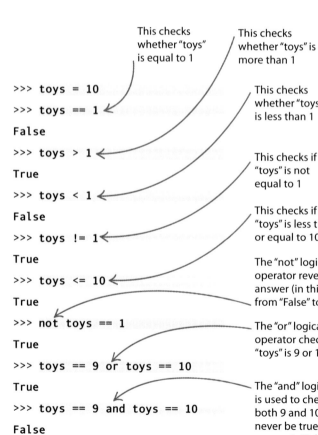

This checks whether "toys" is equal to 1

This checks whether "toys" is more than 1

```
>>> toys = 10
>>> toys == 1
False
>>> toys > 1
True
>>> toys < 1
False
>>> toys != 1
True
>>> toys <= 10
True
>>> not toys == 1
True
>>> toys == 9 or toys == 10
True
>>> toys == 9 and toys == 10
False
```

This checks whether "toys" is less than 1

This checks if "toys" is not equal to 1

This checks if "toys" is less than or equal to 10

The "not" logical operator reverses the answer (in this example, from "False" to "True")

The "or" logical operator checks if "toys" is 9 or 10

The "and" logical operator is used to check if "toys" is both 9 and 10. This can never be true, so the answer is "False"

Is it Ella's birthday?

Ella's birthday is the 28th of July. This program takes a day and a month and uses logical operators to check whether it's Ella's birthday.

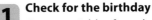
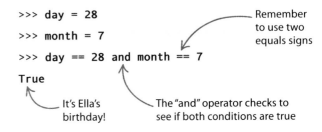

1 Check for the birthday

Create variables for a day and a month. Use the "and" logical operator to check whether it is the 28th of July.

```
>>> day = 28
>>> month = 7
>>> day == 28 and month == 7
True
```

Remember to use two equals signs

It's Ella's birthday!

The "and" operator checks to see if both conditions are true

2 Not the birthday detector

You can reverse the answer using the "not" logical operator. You will get the answer "True" on every day, except for Ella's birthday.

```
>>> day = 28
>>> month = 7
>>> not (day == 28 and \
    month == 7)
False
```

This character is used to make code go over two lines

It's Ella's birthday, so the answer is "False"

3 Birthday or New Year's Day?

Use the "or" logical operator to check whether it's Ella's birthday or New Year's Day. Use brackets to combine the correct days and months.

```
>>> day = 28
>>> month = 7
>>> (day == 28 and month == 7) \
    or (day == 1 and month == 1)
True
```

Checks for the 28th of July.

The answer will be "True" if it's Ella's birthday or New Year's Day

Strings

Two strings can be compared using the "==" operator or the "!=" operator. Strings have to match exactly to get a "True" output.

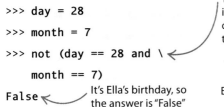

```
>>> dog = "Woof woof"
>>> dog == "Woof woof"
True
>>> dog == "woof woof"
False
>>> dog == "Woof woof "
False
```

The strings match exactly, so the answer is "True"

The strings don't match because there isn't a capital "W"

The strings don't match because there's extra space before the quote mark

△ **Exactly the same**
Strings must match for them to be equal. That means they must use capital letters, spaces, and symbols in exactly the same way.

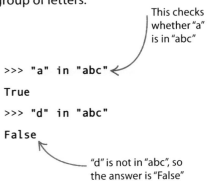

■ ■ **EXPERT TIPS**

Operator for strings

The "in" operator can be used to see whether one string is inside another string. Use it to check if a string contains a particular letter or a group of letters.

This checks whether "a" is in "abc"

```
>>> "a" in "abc"
True
>>> "d" in "abc"
False
```

"d" is not in "abc", so the answer is "False"

Branching

Boolean expressions can be used to determine which route a program should follow, depending on whether the answer to the expression is "True" or "False". This is known as "branching".

SEE ALSO

❮ **64–65** Decisions and branches

❮ **118–119** Making decisions

Do or do not

The "if" command means that if a condition is "True", then the program runs a block of commands. If the condition isn't "True", the block is skipped. The block after the "if" command is always indented by four spaces.

1 **"if" condition**
This code asks the user if it's their birthday. It checks whether the answer is "y". If so, a birthday message is printed.

Indented by four spaces

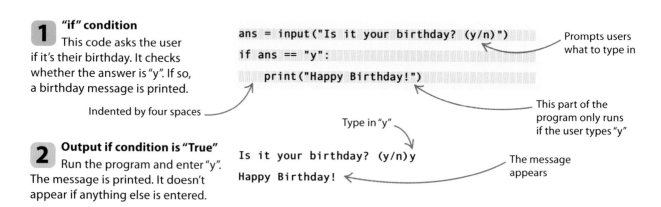

```
ans = input("Is it your birthday? (y/n)")
if ans == "y":
    print("Happy Birthday!")
```

Prompts users what to type in

This part of the program only runs if the user types "y"

2 **Output if condition is "True"**
Run the program and enter "y". The message is printed. It doesn't appear if anything else is entered.

Type in "y"

```
Is it your birthday? (y/n)y
Happy Birthday!
```

The message appears

Do this or that

The "if" command can be combined with an "else" command. This combination means that if something is "True", one thing happens, and if not, something else happens.

1 **"if-else" condition**
If "y" is entered, the program prints a special message for New Year. It shows a different message if anything else is entered.

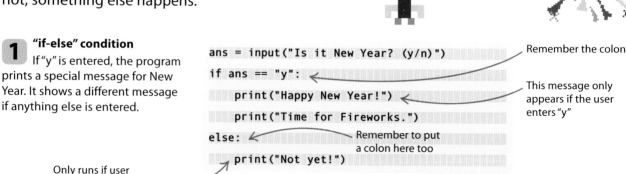

```
ans = input("Is it New Year? (y/n)")
if ans == "y":
    print("Happy New Year!")
    print("Time for Fireworks.")
else:
    print("Not yet!")
```

Remember the colon

This message only appears if the user enters "y"

Remember to put a colon here too

Only runs if user does not enter "y"

2 **Output if condition is "True"**
Run the program and type in "y". The program shows your New Year message. It doesn't show the other message.

```
Is it New Year? (y/n)y
Happy New Year!
Time for Fireworks.
```
Type in "y"

3 **"else" condition output**
Type in "n", or any other character, and the New Year message isn't shown. Instead, the "Not yet!" message appears.

```
Is it New Year? (y/n)n
Not yet!
```
Type in "n"

A different message appears

Do one of these things

The "elif" command is short for "else-if". It means that if something is "True", do one thing, otherwise check if something else is "True" and do something else if it is. The following calculator program uses the "elif" command.

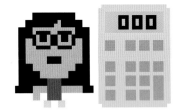

1 **"if-elif-else" condition**
This program checks what is typed in. If it's "add", "sub", "mul", or "div", the result of the sum is shown.

Asks the user to input a number

Remember to add quote marks and brackets

```
a = int(input("a = "))
b = int(input("b = "))
op = input("add/sub/mul/div:")
if op == "add":
    c = a + b
elif op == "sub":
    c = a - b
elif op == "mul":
    c = a * b
elif op == "div":
    c = a / b
else:
    c = "Error"
print("Answer = ",c)
```

Type "add" to add the variables together

Type "div" to divide the variables

Shows the answer or error message

Shows an error message in "c" if something else is typed in

2 **Output for the condition that's "True"**
Test the program. Enter two numbers and type in "sub". The answer will be the first number minus the second number.

```
a = 7
b = 5
add/sub/mul/div:sub
Answer = 2
```
Enter two numbers

Type in "sub" to subtract 5 from 7

Answer is calculated by subtracting variable "a" from variable "b"

3 **"else" condition output**
The "else" condition runs if something other than "add", "sub", "mul", or "div" is typed in, and an error message is displayed.

```
a = 7
b = 5
add/sub/mul/div:try
Answer = Error
```
Type something different here

Error message displays

Loops in Python

Programs that contain repeating lines of code can be time-consuming to type in and difficult to understand. A clearer way of writing them is by using a loop command. The simplest loops are ones that repeat a certain number of times, such as "for" loops.

SEE ALSO

⟨ **48–49** Pens and turtles

While loops **124–125** ⟩

Escaping **126–127** ⟩ loops

Repeating things

A "for" loop repeats the code without having to type it in again. It can be used to repeat something a certain number of times. For example, if you want to print the names of a class of 30 students.

1 **Program the turtle**
A "for" loop can also be used to shorten the code. This program allows the user to control a turtle that draws a line as it moves around the screen. The user can draw shapes on the screen, such as a triangle, by directing the turtle's movements.

This makes the turtle turn 120 degrees to the right

```
from turtle import *
forward(100)
right(120)
forward(100)
right(120)
forward(100)
right(120)
```

Loads all the commands that control the turtle

This command moves the turtle forward

2 **The turtle draws a triangle**
The program tells the turtle how to draw a triangle by giving it the length of the three sides and the angles between them. The turtle will appear in a separate window when you run the program.

The turtle in Python

The program makes the turtle draw a triangle

3 **Use a "for" loop**
The program above gives the turtle the same two commands, "forward(100)" and "right(120)", three times – once for each side of the triangle. An alternative to this is to use these two commands inside a "for" loop. Try drawing a triangle simply using the code shown below.

```
for i in range(3):
    forward(100)
    right(120)
```

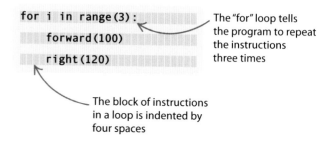

The "for" loop tells the program to repeat the instructions three times

The block of instructions in a loop is indented by four spaces

Loop variables

A loop variable counts the number of times a loop has repeated itself. It starts at the first value in the range (0) and stops one before the last value.

The loop variable

The loop repeats ten times

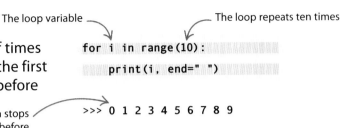

```python
for i in range(10):
    print(i, end=" ")
```

Python stops counting one before the last value

```
>>> 0 1 2 3 4 5 6 7 8 9
```

△ **Simple loop variable**
Here, the loop's range doesn't state what the starting value should be. So Python starts counting from 0, the same way as it does with strings.

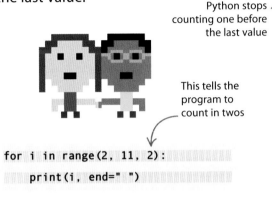

This tells the program to count in twos

```python
for i in range(2, 11, 2):
    print(i, end=" ")
```

```
>>> 2 4 6 8 10
```
The output appears in twos

This tells the program to count backwards

```python
for i in range(10, 0, -1):
    print(i, end=" ")
```

```
>>> 10 9 8 7 6 5 4 3 2 1
```

△ **Counting in twos**
This loop has a third value in its range, which tells the loop to count in twos. It stops at 10, which is one loop before the loop variable gets to 11.

△ **Counting backwards**
This time the program counts backwards from 10, like in a rocket launch. The loop variable starts at 10 and takes steps of -1 until it reaches 1.

Nested Loops

Loops inside a loop are called "nested loops". In nested loops, the outer loop only repeats after the inner loop has gone round its required number of times.

To make the loops repeat "n" number of times, the last number in the range must be "n + 1"

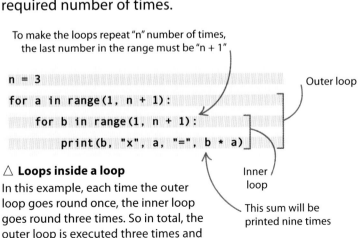

```python
n = 3
for a in range(1, n + 1):
    for b in range(1, n + 1):
        print(b, "x", a, "=", b * a)
```

Outer loop

Inner loop

This sum will be printed nine times

The value of "b"

The value of "a"

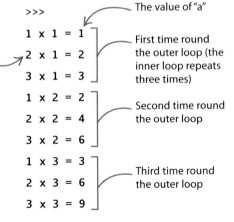

```
>>>
1 x 1 = 1
2 x 1 = 2       First time round
3 x 1 = 3       the outer loop (the
                inner loop repeats
                three times)
1 x 2 = 2
2 x 2 = 4       Second time round
3 x 2 = 6       the outer loop
1 x 3 = 3
2 x 3 = 6       Third time round
3 x 3 = 9       the outer loop
```

△ **Loops inside a loop**
In this example, each time the outer loop goes round once, the inner loop goes round three times. So in total, the outer loop is executed three times and the inner loop is executed nine times.

△ **What happens**
The nested loops print the first three lines of the 1, 2, and 3 times tables. The value of "a" only changes when the outer loop repeats. The value of "b" counts from 1 to 3 for each value of "a".

While loops

"For" loops are useful when you know how many times a task needs to be repeated. But sometimes you'll need a loop to keep repeating until something changes. A "while" loop keeps on going round as many times as it needs to.

SEE ALSO

⟨ **118–119** Making decisions

⟨ **122–123** Loops in Python

Escaping **126–127** ⟩ loops

While loops

A while loop keeps repeating as long as a certain condition is true. This condition is called the "loop condition" and is either true or false.

1 Create a while loop
Set the starting value of the "answer" variable in the loop condition. The loop condition has to be true to start with or the program will never run the loop.

The "answer" variable is set to "y"

The code inside the loop must be indented four spaces.

```
answer = "y"
while answer == "y":
    print("Stay very still")
    answer = input("Is the monster friendly? (y/n)")
print("Run away!")
```

The while loop only runs if the condition is true

If the condition is false, unindented code after the loop runs and a different message appears

▷ **How it works**
A while loop checks if the condition is true. If it is, it goes round the loop again. If it's not, it skips the loop.

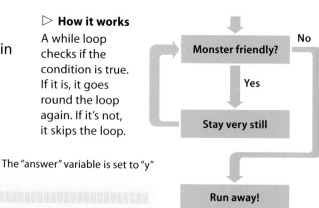

2 What the program looks like
The value entered is stored in the variable "answer". The loop condition is "answer == "y"". If you type "y", the loop keeps going. If you type "n", the loop stops.

```
>>>
Stay very still
Is the monster friendly? (y/n)y
Stay very still
Is the monster friendly? (y/n)y
Stay very still
Is the monster friendly? (y/n)n
Run away!
```

Answer is "y", so the loop keeps running

Answer is "n", so the loop ends and a new message appears

⊡ REMEMBER

"repeat until" block

Python's "while" loop is similar to the "repeat until" block in Scratch. Both keep on repeating until something different happens in the program.

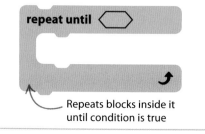

Repeats blocks inside it until condition is true

Forever loops

Some loops run forever. If you set the condition in a "while" loop to be "True", it can never be false and the loop will never end. This can either be useful or very annoying.

△ Going loopy
A loop with the condition "True" is called an "infinite" loop. If something is infinite it has no end.

1 **Create a forever loop**
The loop condition here is set to "True". Nothing that happens inside the loop will make "True" equal anything but "True", so the loop runs forever.

The loop is always "True" so will never end

```
while True:
    answer = input("Type a word and press enter: ")
    print("Please do not type \"" + answer + "\" again.")
```

The typed word is stored in the variable "answer"

2 **What the program looks like**
On the opposite page the monster program's loop condition checked to see what the user's answer was. If the answer isn't "y", the loop will stop. The loop shown above doesn't check the answer, so the user can't make it stop.

```
>>>
Type a word and press enter: tree
Please do not type "tree" again
Type a word and press enter: hippo
Please do not type "hippo" again
Type a word and press enter: water
Please do not type "water": again
Type a word and press enter
```

No matter what is typed, this loop just keeps on going

▪ ▪ ▪ REMEMBER

"forever" block

Remember the "forever" block in Scratch? It repeats the code inside it until the red stop button is clicked. A "while True" loop does exactly the same thing. It can be used to make a program keep doing something, such as asking questions or printing a number, as long as the program is running.

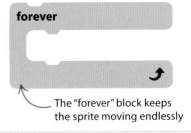

forever

The "forever" block keeps the sprite moving endlessly

▪ ▪ ▪ EXPERT TIPS

Stop the loop

If you get stuck in an infinite loop, you can stop it from IDLE. Click in the Python shell window, then hold down the "CTRL" key and press the "C" key. This asks IDLE to stop the program. You might have to press "CTRL-C" a few times. This is similar to clicking the red stop button in Scratch.

Ctrl-C

Escaping loops

Programs can get stuck in a loop, but there are ways to escape. The word "break" leaves a loop (even a "forever" loop), and the word "continue" skips back to the start of the next loop.

SEE ALSO

❮ **122–123** Loops in Python

❮ **124–125** While loops

Inserting breaks

Putting a break into a loop makes the program jump out of the loop at once – even if the loop condition is true. Any commands inside the loop that come after the break are ignored.

The variable "i" will count from 1 to 12

```
table = 7
for i in range(1, 13):
    print("What's", i, "x", table, "?")
    guess = input()
    ans = i * table
    if int(guess) == ans:
        print("Correct!")
    else:
        print("No, it's", ans)
print("Finished")
```

"i" is the loop variable

1 **Write a simple program**
This program tests the user on the 7 times table. The program continues looping until all 12 questions are answered. Write this program in the code window, as it will be edited later.

2 **Insert a "break"**
A "break" can be added so the user can escape the loop. The program executes a break if the user types "stop".

If "guess" equals "stop", the program skips the rest of the loop and prints "Finished"

```
table = 7
for i in range(1,13):
    print("What's", i, "x", table, "?")
    guess = input()
    if guess == "stop":
        break
    ans = i * table
    if int(guess) == ans:
        print("Correct!")
    else:
        print("No, it's", ans)
print("Finished")
```

The "ans" variable holds the correct answer to the question

```
>>>
What's 1 x 7 ?
1
No, it's 7
What's 2 x 7 ?
14
Correct!
What's 3 x 7 ?
stop
Finished
```

The first time around the loop "i" is equal to 1

The value of "i" changes to 2 next time around the loop

This executes the break command and the program exits the loop

3 How it works
If the user decides not to carry on after the third question and types "stop", the break command is executed and the program leaves the loop.

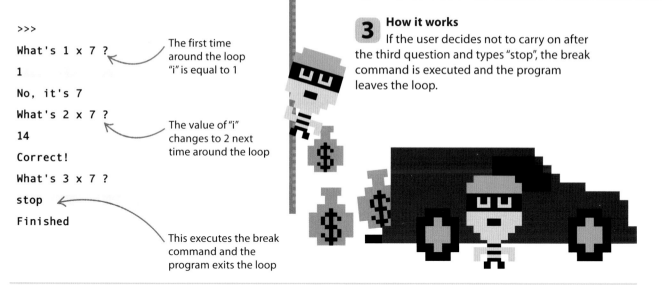

Skipping

The "continue" keyword can be used to skip a question without leaving the loop. It tells the program to ignore the rest of the code inside the loop and skip straight to the start of the next loop.

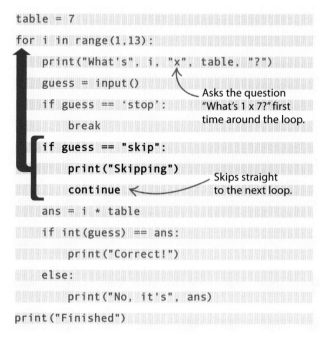

```python
table = 7
for i in range(1,13):
    print("What's", i, "x", table, "?")
    guess = input()
    if guess == 'stop':
        break
    if guess == "skip":
        print("Skipping")
        continue
    ans = i * table
    if int(guess) == ans:
        print("Correct!")
    else:
        print("No, it's", ans)
print("Finished")
```

Asks the question "What's 1 x 7?" first time around the loop.

Skips straight to the next loop.

4 Insert a continue
Add an "if" statement inside the loop to see if the user answered "skip". If so, the program will print "Skipping" and execute a "continue" to skip to the next go around the loop.

5 What happens
If the user doesn't want to answer a question, they can type "skip" and continue to the next question.

```
>>>
What's 1 x 7 ?
skip
Skipping
What's 2 x 7 ?
14
Correct!
What's 3 x 7 ?
```

Type "skip" to go to the next question

The loop goes around again as normal when the answer is correct

Lists

If you need to keep lots of data in one place, then you can put it in a list. Lists can contain numbers, strings, other lists, or a combination of all these things.

SEE ALSO

❮ **54–55** Strings and lists

Silly **132–133** ❯ sentences

What is a list?

A list is a structure in Python where items are kept in order. Each entry is given a number that you can use to refer back to it. You can change, delete, or add to the items in a list at any point.

▽ **Looking at lists**
Each item in a list sits inside double quote marks, and is separated from the next item by a comma. The whole list sits inside a pair of square brackets.

The list is stored in the variable "mylist"

```
>>> mylist = ["apple", "milk", "cheese", "icecream", "lemonade", "tea"]
```

The items in the list sit inside a pair of square brackets

The items in a list are separated by commas

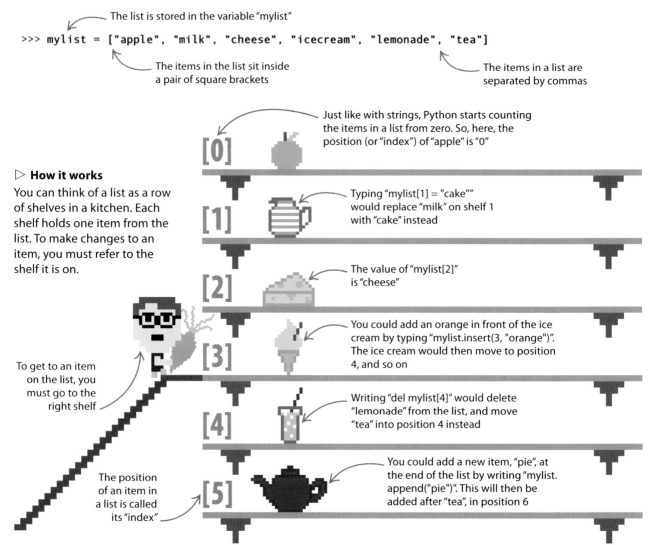

▷ **How it works**
You can think of a list as a row of shelves in a kitchen. Each shelf holds one item from the list. To make changes to an item, you must refer to the shelf it is on.

To get to an item on the list, you must go to the right shelf

The position of an item in a list is called its "index"

[0] Just like with strings, Python starts counting the items in a list from zero. So, here, the position (or "index") of "apple" is "0"

[1] Typing "mylist[1] = "cake"" would replace "milk" on shelf 1 with "cake" instead

[2] The value of "mylist[2]" is "cheese"

[3] You could add an orange in front of the ice cream by typing "mylist.insert(3, "orange")". The ice cream would then move to position 4, and so on

[4] Writing "del mylist[4]" would delete "lemonade" from the list, and move "tea" into position 4 instead

[5] You could add a new item, "pie", at the end of the list by writing "mylist.append("pie")". This will then be added after "tea", in position 6

Using lists

Once a list has been created, you can write programs to manipulate the data inside it – in a loop, for example. You can also combine lists to make new lists.

Mutable objects

Lists in Python are "mutable". This means that they can change. You can add or delete items, or switch around their order. Other functions in Python, such as tuples (see pp.134–135), can't be altered once you create them. These are called "immutable".

The list is stored in the variable "names"

The body of the loop must be indented by four spaces

```
>>> names = ["Simon", "Kate", "Vanya"]
>>> for item in names:
        print("Hello", item)
```

```
Hello Simon
Hello Kate
Hello Vanya
```

When run, this program displays "Hello", followed by each name on the list

◁ **Lists in loops**
You can use a loop to work through every item in a list. This program says "Hello" to a series of names, one after the other.

Remember, lists are contained within square brackets

▷ **Adding lists**
Two lists can be added together. The new list will contain the items from both of the old lists.

```
x = [1, 2, 3, 4]
y = [5, 6, 7, 8]
z = x + y
print(z)
z = [1, 2, 3, 4, 5, 6, 7, 8]
```

This adds the lists together

The new list contains everything from list "x" followed by everything from list "y"

▽ **Lists in lists**
The items in a list can be lists themselves. The "suitcase" list below contains two lists of clothes – it is like a suitcase shared by two people, where they each pack three items.

As the list is inside square brackets, it becomes an individual item within the "suitcase" list – "suitcase[0]"

"suitcase[1]"

```
>>> suitcase=[["hat", "tie", "sock"],["bag", "shoe", "shirt"]]
>>> print(suitcase)
[["hat", "tie", "sock"],["bag", "shoe", "shirt"]]
>>> print(suitcase[1])
["bag", "shoe", "shirt"]
>>> print(suitcase[1][2])
shirt
```

This will print the whole suitcase list

This will print everything in the second list, "suitcase[1]"

This prints the item at index 2 in "suitcase[1]" – remember, Python starts counting the items from zero

Functions

A function is a piece of code that performs a specific task. It bundles up the code, gives it a name, and can be used any time by "calling" it. A function can be used to avoid entering the same lines of code more than once.

SEE ALSO

Silly **132–133 ⟩**
sentences

Variables and **138–139 ⟩**
functions

Useful functions

Python contains lots of useful functions for performing certain tasks. When a function is called, Python retrieves the code for that function and then runs it. When the function is finished, the program returns to the line of code that called it and runs the next command.

print()

△ **"print()" function**
This function lets the program send output to the user by printing instructions or results on the screen.

input()

△ **"input()" function**
This function is the opposite of the "print()" function. It lets the user give instructions or data to the program by typing them in.

randint()

△ **"randint()" function**
This function gives a random number (like throwing a dice). It can be used to add an element of chance to programs.

Making and calling functions

The functions that come with Python aren't the only ones that can be used. To make a new function, collect the code you want to use in a special "wrapper" and give it a name. This name allows the function to be called whenever it is needed.

1 **Define a function**
The definition of a function will always have the keyword "def" and the function's name at the beginning of the code.

```
def greeting():
    print("Hello!")
```

A colon marks the end of the function's name and the start of the code it contains

This is the code within the function

2 **Call the function**
Typing the function name followed by brackets into the shell window calls the function and shows the output.

```
>>> greeting()
Hello!
```

The "greeting" function is called and the output is displayed

Brackets show that this is a function call and not a variable

Passing data to functions

A function has to be told which values to work with.
For example, in "print(a, b, c)", the function "print()" is being
passed the values "a", "b", and "c". In "height(1, 45)", the values
1 and 45 are being passed to the function "height".

1 Add parameters to the function
Values passed to a function are called
"parameters". Parameters are put inside the brackets
next to the function's name in its definition.

"m" and "cm" are the parameters

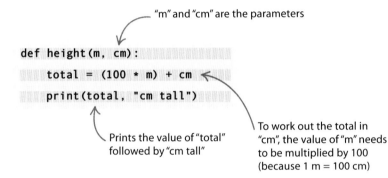

```
def height(m, cm):
    total = (100 * m) + cm
    print(total, "cm tall")
```

Prints the value of "total"
followed by "cm tall"

To work out the total in
"cm", the value of "m" needs
to be multiplied by 100
(because 1 m = 100 cm)

2 Values are defined
The code inside the function
uses the values that are passed to it.

Calls the function to give
the answer when "m" = 1
and "cm" = 45

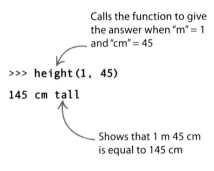

```
>>> height(1, 45)
145 cm tall
```

Shows that 1 m 45 cm
is equal to 145 cm

Getting data back from functions

Functions are most useful when they send some data
back to the program – a return value. To make a
function return a value, add "return" followed by the
value to be sent back.

1 Define a function that returns a number
Python's "input()" function always returns a string,
even if a number is entered. The new function below
gives back a number instead.

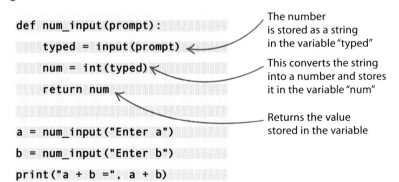

```
def num_input(prompt):
    typed = input(prompt)
    num = int(typed)
    return num

a = num_input("Enter a")
b = num_input("Enter b")
print("a + b =", a + b)
```

The number
is stored as a string
in the variable "typed"

This converts the string
into a number and stores
it in the variable "num"

Returns the value
stored in the variable

2 Number as output
If the program used the function
"input", "a + b" would put the strings "10"
and "7" together to give "107".

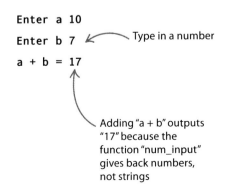

```
Enter a 10
Enter b 7
a + b = 17
```

Type in a number

Adding "a + b" outputs
"17" because the
function "num_input"
gives back numbers,
not strings

> **PROJECT 5**

Silly sentences

Loops, functions, and lists can be used individually for lots of different tasks. They can also be used together to create interesting programs that can do even more complex tasks.

SEE ALSO

❮ **124–125** While loops

❮ **128–129** Lists

❮ **130–131** Functions

Make silly sentences

This program will make sentences by using three separate lists of words. It will pick one word from each list and put them together randomly in a silly sentence.

Try using different words to the ones shown here to create your own silly sentences.

1 Enter the three lists shown below into a new code window. This defines the lists that will be used to make the sentences.

Double quotes show that each item in the list is a string

```
name = ["Neha", "Lee", "Sam"]
verb = ["buys", "rides", "kicks"]
noun = ["lion", "bicycle", "plane"]
```

Square brackets mean that this is a list

2 Each sentence is made up of words picked at random from the lists you have created. Define a function to do this, as it will be used several times in the program.

This loads the function for generating a random number ("randint")

Finds out how many words are in the list (the function works for lists of any length)

```
from random import randint
def pick(words):
    num_words = len(words)
    num_picked = randint(0, num_words - 1)
    word_picked = words[num_picked]
    return word_picked
```

Picks a random number that refers to one of the items in the list

Stores the random word that has been picked in the variable "word_picked"

3 Print a random silly sentence by running the "pick" function once for each of the three lists. Use the "print" command to show the sentence on the screen.

This adds a full stop at the end, while the "\n" starts a new line

```
print(pick(name), pick(verb), "a", pick(noun), end=".\n")
```

Add an "a" so that the sentence makes sense (see below)

4 Save and run the program to get a silly sentence made from the lists of names, verbs, and nouns.

```
Neha kicks a bicycle.
```

The sentence is randomly selected each time the program is run

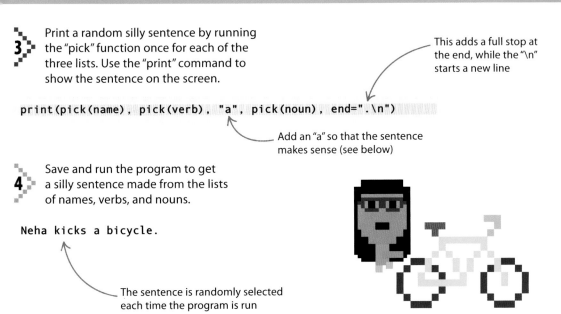

Silly sentences forever!

A forever loop can be added to the silly sentences program to keep it running forever, or until the user presses "Ctrl-C" to escape the loop.

1 The program keeps printing silly sentences if the "print" command is wrapped in a "while True" loop.

Wraps the print command in a loop

```
while True:
    print(pick(name), pick(verb), "a", pick(noun), end=".")
    input()
```

Prints a new sentence every time the "Enter" key is pressed

The program will keep on creating random sentences

2 The "input()" function waits for the user to press the "Enter" key before printing another sentence. Without this it would print them too fast to read.

```
Sam rides a lion.
Neha kicks a plane.
Lee buys a bicycle.
```

> ■ ■ **EXPERT TIPS**
> ## Readable code
>
> It's very important to write a program that can be easily understood. It makes the program easier to change in the future because you don't have to start by solving the puzzle of how it works!

Tuples and dictionaries

Python uses lists for keeping data in order. It also has other data types for storing information called "tuples" and "dictionaries". Data types such as these, which hold lots of items, are called "containers".

SEE ALSO

⟨ **110–111** Types of data

⟨ **128–129** Lists

Tuples

Tuples are a bit like lists, but the items inside them can't be changed. Once a tuple is set up it always stays the same.

Tuples are surrounded by brackets

```
>>> dragonA = ("Sam", 15, 1.70)
>>> dragonB = ("Fiona", 16, 1.68)
```

The items in a tuple are separated by commas

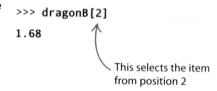

◁ **What is a tuple?**
A tuple contains items separated by commas and surrounded by brackets. Tuples are useful for collecting several bits of data together, such as a dragons' name, age, and height.

▷ **Grabbing an item from a tuple**
To get an item from a tuple, use its position in the tuple (its index). Tuples count from zero, just like lists and strings.

```
>>> dragonB[2]
1.68
```

This selects the item from position 2

```
>>> name, age, height = dragonA
>>> print(name, age, height)
Sam 15 1.7
```

The items that make up the tuple "dragonA" are displayed separately

◁ **Splitting a tuple into variables**
Assign three variables to the tuple "dragonA" – "name", "age", and "height". Python splits the tuple into three items, putting one in each variable.

▷ **Putting tuples in a list**
Tuples can be put into a list because containers can go inside each other. Use this code to create a list of tuples.

Create a list of tuples called "dragons"

Lists go in square brackets

```
>>> dragons = [dragonA, dragonB]
>>> print(dragons)
[("Sam", 15, 1.7), ("Fiona", 16, 1.68)]
```

Each tuple is surrounded by round brackets inside the list's square brackets

Python displays all the items that are in the list, not just the names of the tuples

Dictionaries

Dictionaries are like lists but they have labels. These labels, called "keys", identify items instead of index numbers. Every item in a dictionary has a key and a value. Items in a dictionary don't have to stay in a particular order, and the contents of a dictionary can be changed.

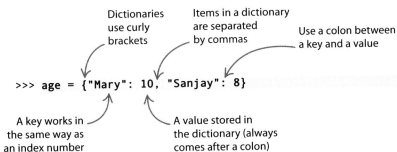

Dictionaries use curly brackets

Items in a dictionary are separated by commas

Use a colon between a key and a value

▷ **Create a dictionary**

This program creates a dictionary called "age". The key for each item is the name of a person. The value is their age.

```
>>> age = {"Mary": 10, "Sanjay": 8}
```

A key works in the same way as an index number

A value stored in the dictionary (always comes after a colon)

```
>>> print(age)
{"Sanjay": 8, "Mary": 10}
```

Name of the dictionary

The key for this item is "Sanjay"

The value of "Mary" is 10

◁ **Print the dictionary**

The order of the items can change, because the positions of items in a dictionary are not fixed.

Dictionary name

New key

▷ **Add a new item**

A new value can be added to the dictionary by labelling it with the new key.

```
>>> age["Owen"] = 11
>>> print(age)
{"Owen": 11, "Sanjay": 8, "Mary": 10}
```

Adds a new item to the dictionary

The new value is now in the dictionary

The existing values are still there

Assign a new value to the item labelled "Owen"

```
>>> age["Owen"] = 12
>>> print(age)
{"Owen": 12, "Sanjay": 8, "Mary": 10}
```

◁ **Change a value**

Assign a new value to an existing key to change its value.

The value for "Owen" has changed

This deletes the item labelled "Owen"

▷ **Delete an item**

Deleting an item in a dictionary doesn't affect other items because they are identified by their key, not by their position in the dictionary.

```
>>> del age["Owen"]
>>> print(age)
{"Sanjay": 8, "Mary": 10}
```

The item labelled "Owen" no longer appears in the dictionary

Lists in variables

There's something about how Python stores lists in variables that might seem a bit odd at first. But take a look at what's going on behind the scenes and it all makes sense.

SEE ALSO

❮ **108–109** Variables in Python

❮ **128–129** Lists

Remember how variables only store values?

Variables are like boxes that hold values. The value in one variable can be copied and stored in another. It's like photocopying the value contained in box "a" and storing a copy in box "b".

△ **How variables work**
Each variable is like a box containing a piece of paper with a value written on it.

1 **Assign a value to a variable**
Assign the value 2 to variable "a", then assign the value in "a" to variable "b". The value 2 is copied and stored in "b".

This copies the contents of "a" into "b"

```
>>> a = 2
>>> b = a
>>> print("a =", a, "b =", b)
a = 2 b = 2
```

This prints out the variable names with their values

Now "a" and "b" both contain the value 2

2 **Change a value**
If you change the value stored in one variable it won't affect the value stored in another variable. In the same way changing what's written on a piece of paper in box "a" won't affect what's on the paper in box "b".

Change the value in "a" to 100

```
>>> a = 100
>>> print("a =", a, "b =", b)
a = 100 b = 2
```

Now "a" contains 100, but "b" still contains 2

3 **Change a different value**
Change the value in "b" to 22. Variable "a" still contains 100. Even though the value of "b" was copied from "a" at the start, they are now independent – changing "b" doesn't change "a".

```
>>> b = 22
>>> print("a =", a, "b =", b)
a = 100 b = 22
```

"b" now contains 22, but "a" is still 100

What happens if a list is put in a variable?

Copying the value in a variable creates two independent copies of the value. This works if the value is a number, but what about other types of value? If a variable contains a list it works a bit differently.

1 Copy a list
Store the list [1, 2, 3] in a variable called "listA". Then store the value of "listA" in another variable called "listB". Now both variables contain [1, 2, 3].

Use square brackets to create a list

```
>>> listA = [1, 2, 3]
>>> listB = listA
>>> print("listA =", listA, "listB =", listB)
listA = [1, 2, 3] listB = [1, 2, 3]
```

This prints out the variable names alongside their values to see what's inside them

"listA" and "listB" both hold the same value

This changes the second item in the list because lists count from zero

2 Change list A
Change the value in "listA[1]" to 1000. "listB[1]" now contains 1000 as well. Changing the original list has changed the copy of the list too.

```
>>> listA[1] = 1000
>>> print("listA =", listA, "listB =", listB)
listA = [1, 1000, 3] listB = [1, 1000, 3]
```

The second item of both "listA" and "listB" has been changed

This is the third item in the list

3 Change list B
Change the value of "listB[2]" to 75. "listA[2]" is now 75 as well. Changing the copy of the list has changed the original list as well.

```
>>> listB[2] = 75
>>> print("listA =", listA, "listB =", listB)
listA = [1, 1000, 75] listB = [1, 1000, 75]
```

The third item of both "listA" and "listB" has been changed

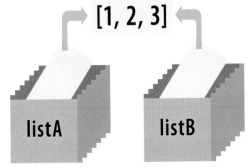

[1, 2, 3]

listA   listB

△ **What's going on?**
A variable containing a list doesn't hold the list itself, just a link to it. Copying the value in "listA" copies the link. So both "listA" and "listB" contain a link to the same list.

Variables and functions

SEE ALSO

❰ **130–131** Functions

Making **158–159** ❱
shapes

Variables created inside a function (local variables) and variables created in the main program (global variables) work in different ways.

Local variables are like film stars in a car with mirrored windows – they are inside the car (function) but no one can see them.

Local variables

Local variables only exist inside a single function, so the main program and other functions can't use them. If you try to use a local variable outside of the function, an error message appears.

1 **Variable inside the function**
Create a local variable called "a" inside "func1". Print out the value of "a" by calling "func1" from the main program.

```
>>> def func1():
        a = 10
        print(a)
>>> func1()
10
```
Calling "func1" prints the value given to "a"

2 **Variable outside the function**
If you try to print "a" directly from the main program, it gives an error. "a" only exists inside "func1".

```
>>> print(a)
Traceback (most recent call last):
  File "<pyshell#6>", line 1, in <module>
    print(a)
NameError: name "a" is not defined
```

The main program doesn't know what "a" is, so it prints an error message

Global variables

A variable created in the main program is called a global variable. Other functions can read it, but they can't change its value.

Global variables are like people walking along the street – everyone can see them

1 **Variable outside the function**
Create a global variable called "b" in the main program. The new function ("func2") can read the value of "b" and print it.

```
>>> b = 1000
>>> def func2():
        print(b)
>>> func2()
1000
```
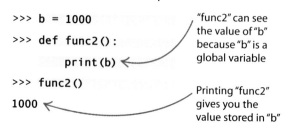

"func2" can see the value of "b" because "b" is a global variable

Printing "func2" gives you the value stored in "b"

2 **Same global variable**
We can also print "b" directly from the main program. "b" can be seen everywhere because it wasn't created inside a function.

```
>>> print(b)
1000
```
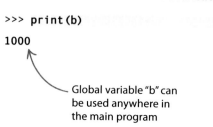

Global variable "b" can be used anywhere in the main program

Variables as input to functions

When a variable is used as input to a function its value is copied into a new local variable. So changing the value of this new local variable inside the function doesn't change the value of the original variable.

1 Changing values inside a variable
"func3" uses input "y", which is a local variable. It prints the value of "y", then changes that value to "bread" and prints the new value.

2 Print variable
Printing the value of "z" after calling "func3" shows it hasn't changed. Calling "func3" copies the value in "z" ("butter") into local variable "y", but "z" is left unchanged.

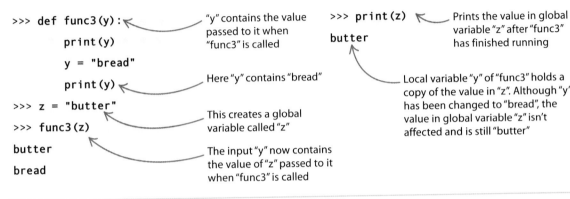

```
>>> def func3(y):
        print(y)
        y = "bread"
        print(y)
>>> z = "butter"
>>> func3(z)
butter
bread
```

"y" contains the value passed to it when "func3" is called

Here "y" contains "bread"

This creates a global variable called "z"

The input "y" now contains the value of "z" passed to it when "func3" is called

```
>>> print(z)
butter
```

Prints the value in global variable "z" after "func3" has finished running

Local variable "y" of "func3" holds a copy of the value in "z". Although "y" has been changed to "bread", the value in global variable "z" isn't affected and is still "butter"

Masking a global variable

A global variable can't be changed by a function. A function trying to change a global variable actually creates a local variable with the same name. It covers up, or "masks", the global variable with a local version.

1 Changing a global variable
Global variable "c" is given the value 12345. "func4" gives "c" the value 555 and prints it out. It looks like our global variable "c" has been changed.

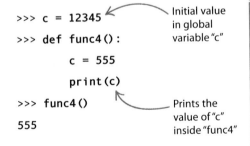

```
>>> c = 12345
>>> def func4():
        c = 555
        print(c)
>>> func4()
555
```

Initial value in global variable "c"

Prints the value of "c" inside "func4"

2 Print variable
If we print "c" from outside the function, we see that "c" hasn't changed at all. "func4" only prints the value of its new local variable – also called "c".

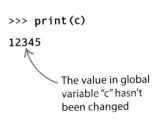

```
>>> print(c)
12345
```

The value in global variable "c" hasn't been changed

⊕ PROJECT 6

Drawing machine

It's time to try a more complex project. This program, the drawing machine, turns a string of simple instructions into turtle commands to draw different shapes. The skills used in planning this program are essential for any coder.

SEE ALSO

❮ **122–123** Loops in Python

Libraries **152–153** ❯

Choose a test shape

To write a program that can draw any shape, it's useful to choose a shape to start with. Use this house shape as an example to test the program at each stage. By the end of the project it will be possible to draw this house with far less code – by using a single string containing several short drawing commands (for example, "F100").

```
from turtle import *
reset()
left(90)
forward(100)
right(45)
forward(70)
right(90)
forward(70)
right(45)
forward(100)
right(90)
forward(100)
```

Loads all the commands that control the turtle

Resets the turtle's position and puts the pen down ready to draw

Moves the turtle forward by 70

Makes the turtle turn 90 degrees to the right

▷ **Turtle draws a house**
The arrow shows the final direction and position of the turtle. Starting at the bottom left, it has moved clockwise around the house.

The turtle

△ **Program to draw a house**
This code tells the turtle to draw a house. It requires lots of lines of code for what is actually quite a simple program.

Three parts of the program

The drawing machine will be a large program. To help with the planning, it can be broken down into three parts, each one related to a different task.

Function 1

△ **Turtle controller**
This function takes a simple command from the user and turns it into a turtle command. The user command will come as a single letter and a number.

Function 2

△ **String artist**
In this program, the user enters a string of instructions. This function splits the string into smaller units, which are then fed to the Turtle controller.

Main program

△ **User interface**
The String artist needs to get its input from somewhere. The User interface allows the user to type in a string of commands for the String artist to work on.

Draw a flowchart

Coders often plan programs on paper, to help them write better code with fewer errors. One way to plan is to draw a flowchart – a diagram of the steps and decisions that the program needs to follow.

1 This flowchart shows the plan for the Turtle controller function. It takes a letter (input "do") and number (input "val") and turns them into a turtle command. For example, "F" and "100" will be turned into the command "forward(100)". If the function doesn't recognize the letter, it reports an error to the user.

Each command has two variables: "do" (a string) tells the turtle what to do, and "val" (an integer, or whole number) tells the turtle how much or how far to do it

The function has to decide if the "do" value is a letter it recognizes

If "do" isn't F, the function runs through other letters it recognizes

"do" isn't "R". Is it "U"?

inputs – do and val

If "do" = F, the turtle moves forward

do == F? — Y → **forward(val)**

If "do" = R, the turtle turns right

do == R? — Y → **right(val)**

do == U? — Y → **penup()**

N

report unknown command

If "do" isn't a letter the function recognizes, it reports an error

Because "do" is "U", the command "penup()" stops the turtle from drawing

return from function

Once the command is finished you return to the main program

After any command is executed successfully, the program goes to the end of the function

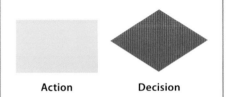

⊙ DRAWING MACHINE

The Turtle controller

The first part of the program is a function that moves the turtle, one command at a time. It is planned out in the flowchart on the previous page. This code enables the turtle to convert the "do" and "val" values into movement commands.

Loads all the commands that control the turtle

2 This code creates the Turtle controller function. It turns "do" inputs into directions for the turtle, and "val" inputs into angles and distances.

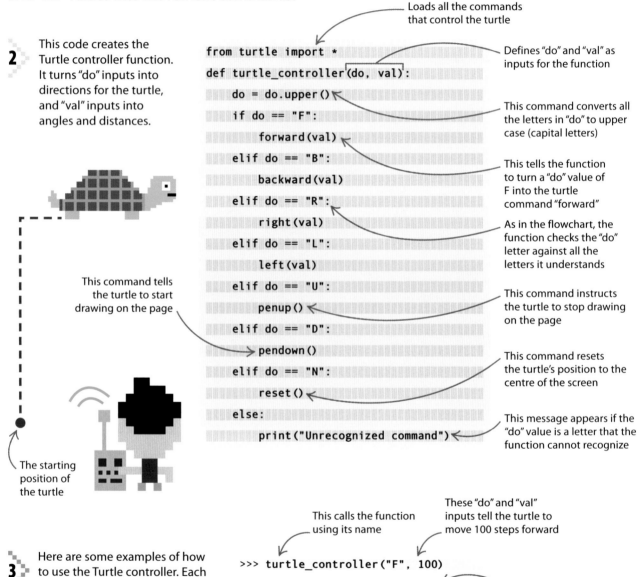

```
from turtle import *
def turtle_controller(do, val):
    do = do.upper()
    if do == "F":
        forward(val)
    elif do == "B":
        backward(val)
    elif do == "R":
        right(val)
    elif do == "L":
        left(val)
    elif do == "U":
        penup()
    elif do == "D":
        pendown()
    elif do == "N":
        reset()
    else:
        print("Unrecognized command")
```

Defines "do" and "val" as inputs for the function

This command converts all the letters in "do" to upper case (capital letters)

This tells the function to turn a "do" value of F into the turtle command "forward"

As in the flowchart, the function checks the "do" letter against all the letters it understands

This command instructs the turtle to stop drawing on the page

This command resets the turtle's position to the centre of the screen

This message appears if the "do" value is a letter that the function cannot recognize

This command tells the turtle to start drawing on the page

The starting position of the turtle

3 Here are some examples of how to use the Turtle controller. Each time it is used, it takes a "do, val" command and turns it into code the turtle can understand.

This calls the function using its name

These "do" and "val" inputs tell the turtle to move 100 steps forward

```
>>> turtle_controller("F", 100)
>>> turtle_controller("R", 90)
>>> turtle_controller("F", 50)
```

This makes the turtle turn right 90 degrees

Write some pseudocode

Another way to plan a program is to write it in pseudocode. "Pseudo" means fake, so pseudocode isn't real code that you can run. It's rough code where you can write your ideas in the style of the real thing.

 4 It's time to plan the String artist. This function takes a string of several "do" and "val" inputs and breaks it into pairs made up of a letter and a number. It then passes the pairs to the Turtle controller one at a time.

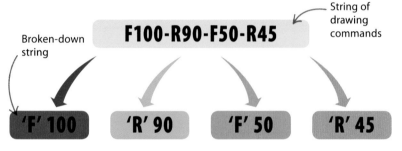

String of drawing commands

F100-R90-F50-R45

Broken-down string

'F' 100 'R' 90 'F' 50 'R' 45

 5 This is the String artist written in pseudocode. It lets you organize the ideas and structure of the code without having to think about the details yet.

function string_artist(input – the program as a string):

The function will take in a string of commands input by the user (for example, "F100-R90")

split program string into list of commands

Splits string into a list of separate commands

for each command in list:

check it's not blank

A blank command won't work, so the function skips it

– if it is go on to next item in list

command type is the first letter

Recognizes the first letter as a "do" command

if followed by more characters

– turn them into a number

Recognizes the following characters as a "val" number

call turtle_controller(command type, number)

Passes the simple command to Turtle controller

⊙ DRAWING MACHINE

Creating the String artist

The pseudocode on the previous page plans a function called the String artist, which will turn a string of values into single commands that are sent to the Turtle controller. The next stage is to turn the pseudocode into real Python code, using a function called "split()".

6 The "split()" function splits a string into a list of smaller strings. Each break point is marked by a special character ("-" in this program).

This string lists the commands to create the sample house shape

```
>>> program = "N-L90-F100-R45-F70-R90-F70-R45-F100-R90-F100"
>>> cmd_list = program.split("-")
>>> cmd_list
["N", "L90", "F100", "R45", "F70", 'R90', "F70", "R45", "F100", "R90", "F100"]
```

The "split()" function breaks the string down into a list of separate commands

7 Now write out the pseudocode for the String artist using real Python code. Use the "split()" function to slice up the input string into turtle commands.

Tells the program to split the string wherever it sees a "-" character

This makes the program loop through the list of strings – each item is one command for the turtle

Gets the length of the command string

Checks if the command is followed by more characters (the number)

Converts the characters from strings into numbers

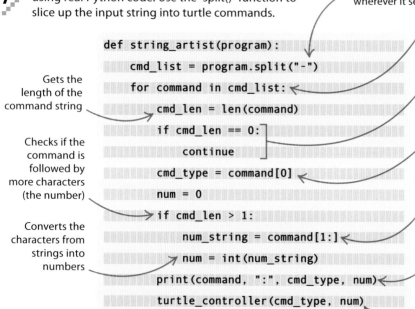

```
def string_artist(program):
    cmd_list = program.split("-")
    for command in cmd_list:
        cmd_len = len(command)
        if cmd_len == 0:
            continue
        cmd_type = command[0]
        num = 0
        if cmd_len > 1:
            num_string = command[1:]
            num = int(num_string)
        print(command, ":", cmd_type, num)
        turtle_controller(cmd_type, num)
```

If the length of the command is 0 (so the command is blank), the function skips it and moves to the next one

Takes the first character of the command (remember, strings start at 0) and sets it as the command type ("F", "U", etc.)

This takes all the remaining characters from the command by cutting off the first one

Prints the command on the screen so you can see what the code is doing

Passes the command to the turtle

8 When the string representing the instructions for the house shape is passed into the String artist, it shows this output in the shell window.

```
>>> string_artist("N-L90-F100-R45-F70-R90-F70-R45-F100-R90-F100")
```

The turtle commands are all separated by a "-"

N : N 0 — Resets the screen and puts the turtle back at the centre

L90 : L 90

F100 : F 100

R45 : R 45 — For command "F100", the command type is "F" and "num" is "100"

F70 : F 70

R90 : R 90 — This makes the turtle turn 45 degrees before drawing the roof

F70 : F 70

R45 : R 45 — This command makes the turtle draw the right-hand side of the roof

F100 : F 100

R90 : R 90 — The turtle turns 90 degrees right, ready to draw the bottom of the house

F100 : F 100

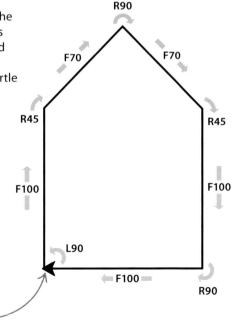

9 Each command in the string that is passed to the "string_artist" function is extracted, identified, and executed. A picture of a house is drawn in the turtle graphics window.

R90

F70   F70

R45   R45

F100   F100

L90

F100

R90

The program makes the turtle draw a house

⏩ DRAWING MACHINE

Finish off the code with a user interface

The drawing machine needs an interface to make it easier to use. This will let the user enter a string from the keyboard to tell the machine what to draw.

10 This code creates a pop-up window where the user can input instructions. A "while True" loop lets them keep entering new strings.

The triple quote (''') tells Python that everything until the next triple quote is part of the same string, including the line breaks

```
instructions = '''Enter a program for the turtle:
eg F100-R45-U-F100-L45-D-F100-R90-B50
N = New drawing
U/D = Pen Up/Down
F100 = Forward 100
B50 = Backwards 50
R90 = Right turn 90 deg
L45 = Left turn 45 deg'''
screen = getscreen()
while True:
    t_program = screen.textinput("Drawing Machine", instructions)
    print(t_program)
    if t_program == None or t_program.upper() == "END":
        break
    string_artist(t_program)
```

Tells the user what letters to use for different turtle commands

End of the string

Gets the data needed to create the pop-up window

This line tells the program what to show in the pop-up window

Stops the program if the user types "END" or presses the "Cancel" button

Passes the string to the String artist function

11 This window pops up over the turtle window ready for the user to type a drawing machine program string.

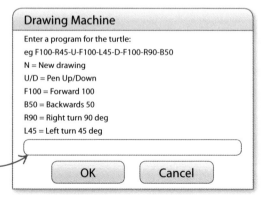

Drawing Machine

Enter a program for the turtle:
eg F100-R45-U-F100-L45-D-F100-R90-B50
N = New drawing
U/D = Pen Up/Down
F100 = Forward 100
B50 = Backwards 50
R90 = Right turn 90 deg
L45 = Left turn 45 deg

OK   Cancel

Type the program string here and then click "OK" to run the program

△ **Turtle control**
Using this program, the turtle is easier to control, and you don't have to restart the program to draw another picture.

12

The drawing machine can be used to create more than just outlines. By lifting up the turtle's pen while moving to a new position, it's possible to fill in details inside a shape. Run the program and try entering the string below.

Lifts up the turtle's pen so it moves without leaving a line

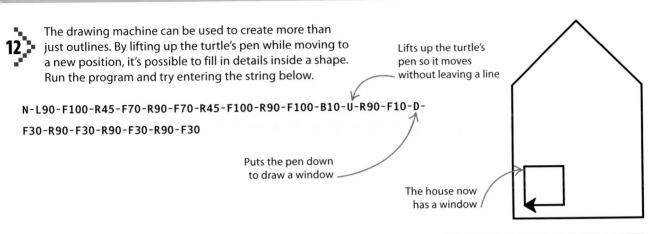

`N-L90-F100-R45-F70-R90-F70-R45-F100-R90-F100-B10-U-R90-F10-D-`

`F30-R90-F30-R90-F30-R90-F30`

Puts the pen down to draw a window

The house now has a window

Time for something different

Now you know how to add details, you can really have fun with the drawing machine. Try drawing this owl face using the string of instructions below.

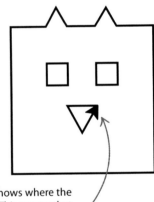

`N-F100-L90-F200-L90-F50-R60-F30-L120-F30-R60-F40-R60-F30-L120-F30-`

`R60-F50-L90-F200-L90-F100-L90-U-F150-L90-F20-D-F30-L90-F30-L90-F30-`

`L90-F30-R90-U-F40-D-F30-R90-F30-R90-F30-R90-F30-L180-U-F60-R90-D-`

`F40-L120-F40-L120-F40`

The string lifts the pen three times to draw the eyes and nose separately

The arrow shows where the turtle stopped. This means that the owl's nose was drawn last

Achievements

You created the drawing machine program by achieving several smaller targets:

Used a flowchart to plan a function by working out the decision points and the resulting actions.

Wrote pseudocode to plan out a function before writing out the real code.

Created the function "turtle_controller" that works out what turtle command to execute from the letter and number it's been given.

Created the function "string_artist" that produced a turtle drawing from a string of instructions.

Made an interface that allows the user to tell the program what to draw from the keyboard.

Bugs and debugging

Programmers aren't perfect, and most programs contain errors at first. These errors are known as "bugs" and tracking them down is called "debugging".

SEE ALSO

❮ **94–95** Errors

❮ **122–123** Loops in Python

What next? **176–177** ❯

Types of bugs

Three main types of bugs can turn up in programs – syntax, runtime, and logic errors. Some are quite easy to spot, while others are more difficult, but there are ways of finding and fixing them all.

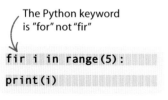

The Python keyword is "for" not "fir"

```
fir i in range(5):
    print(i)
```

This will cause an error as no number can be divided by 0

```
a = 0
print(10 / a)
```

Age cannot be less than 5 and greater than 8 at the same time, so no free tickets

```
if age < 5 and age > 8:
    print("Free ticket!")
```

△ **Easy to spot**

A syntax error is a mistake in the program's words or symbols, such as misspelled keywords, missing brackets, or incorrect indents.

△ **Harder to spot**

Runtime errors appear only when the program is running. Adding numbers to strings or dividing by 0 can cause them.

△ **Hardest to spot**

Logic errors are mistakes in a program's thinking. Using "<" instead of ">", for example, or adding when you should be subtracting result in these errors.

Find and fix a bug

Syntax errors are easy to spot as IDLE highlights them in red when you run the program. Finding runtime and logic errors takes a bit more work.

1 **Problem program**

This program aims to add all the numbers from 1 up to the value stored in the variable "top_num". It then prints the total.

```
top_num = 5
total = 0
for n in range(top_num):
    total = total + n
print("Sum of numbers 1 to", top_num, "is", total)
```

The highest number in the series of numbers being added

This command prints a sentence to let the user know the result

2 **Output**

The answer for the program should be (1 + 2 + 3 + 4 + 5), but it shows the answer as "10". You need to find out why.

```
Sum of numbers 1 to 5 is 10
```

The answer should be "15", not "10"

3 Add a "print" and "input()"

The program doesn't show what it's doing at each step. Adding a "print" command here will let you see what's happening. The "input()" command waits for the "return" or "Enter" key to be pressed before looping.

```
top_num = 5
total = 0
for n in range(top_num):
    total = total + n
    print("DEBUG: n=", n, "total=", total)
    input()
print("Sum of numbers 1 to", top_num, "is", total)
```

This command prints the current value of the loop variable and the total so far

4 New output

The loop is only adding the numbers from 0 up to 4, and not 1 to 5. This is because a "for" loop always starts counting from 0 (unless told otherwise), and always stops 1 before the end of the range.

```
DEBUG: n= 0 total= 0
DEBUG: n= 1 total= 1
DEBUG: n= 2 total= 3
DEBUG: n= 3 total= 6
DEBUG: n= 4 total= 10
Sum of numbers 1 to 5 is 10
```

This is actually the sum of the numbers from 0 to 4, not 1 to 5

5 Fix the faulty line

The range should go from 1 up to "top_num + 1", so that the loop adds up the numbers from 1 to "top_num" (5).

```
top_num = 5
total = 0
for n in range(1, top_num + 1):
    total = total + n
    print("DEBUG: n=", n, "total=", total)
    input()
print("Sum of numbers 1 to", top_num, "is", total)
```

The new range will count from 1 and stop at "top_num" (1 less than "top_num + 1")

6 Correct output

The "print" command shows that the program is adding the numbers from 1 to 5 and getting the correct answer. The bug has now been fixed!

```
DEBUG: n= 1 total= 1
DEBUG: n= 2 total= 3
DEBUG: n= 3 total= 6
DEBUG: n= 4 total= 10
DEBUG: n= 5 total= 15
Sum of numbers 1 to 5 is 15
```

When "n= 3", the total is (1 + 2 + 3)

The correct answer is now printed

Algorithms

An algorithm is a set of instructions for performing a task. Some algorithms are more efficient than others and take less time or effort. Different types of algorithms can be used for simple tasks such as sorting a list of numbers.

SEE ALSO

❮ **16–17** Think like a computer

Libraries **152–153** ❯

Insertion sort

Imagine you've been given your class's exam papers to put in order from the lowest to the highest mark. "Insertion sort" creates a sorted section at the top of the pile and then inserts each unsorted paper into the correct position.

△ **Sorting in order**
"Insertion sort" takes each paper in turn and inserts it into the correct (sorted) place.

▽ **How it works**
"Insertion sort" goes through each of these stages sorting the numbers far quicker than a human could.

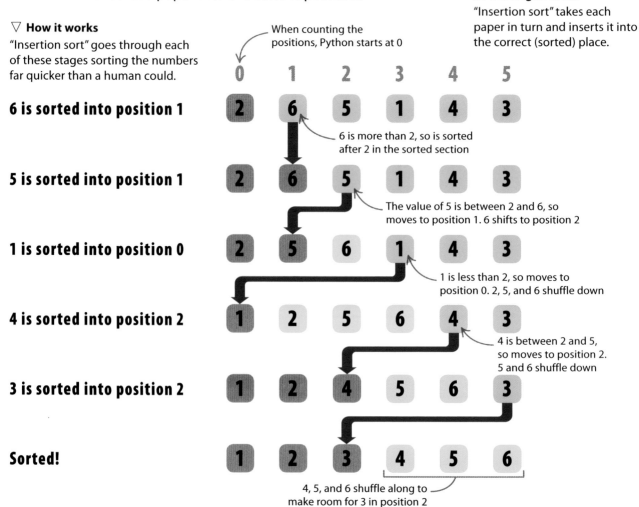

When counting the positions, Python starts at 0

6 is sorted into position 1

6 is more than 2, so is sorted after 2 in the sorted section

5 is sorted into position 1

The value of 5 is between 2 and 6, so moves to position 1. 6 shifts to position 2

1 is sorted into position 0

1 is less than 2, so moves to position 0. 2, 5, and 6 shuffle down

4 is sorted into position 2

4 is between 2 and 5, so moves to position 2. 5 and 6 shuffle down

3 is sorted into position 2

Sorted!

4, 5, and 6 shuffle along to make room for 3 in position 2

Selection sort

"Selection sort" works differently to "insertion sort". It swaps pairs of items rather than constantly shifting all of the items. Each swap moves one number to its final (sorted) position.

△ **Swapping positions**
Switching one thing with another is usually quick and doesn't affect anything else in the list.

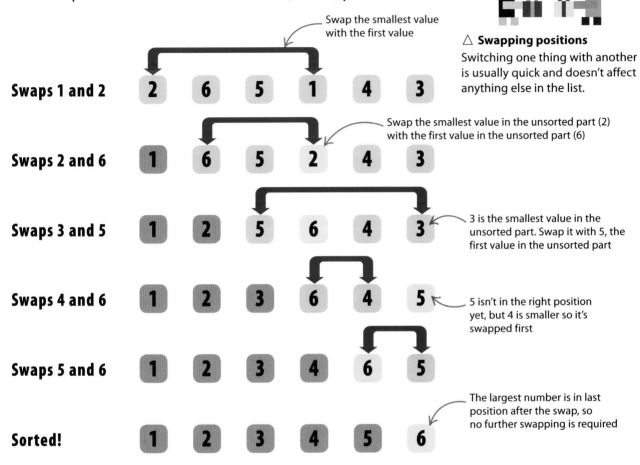

Swap the smallest value with the first value

Swaps 1 and 2 2 6 5 1 4 3

Swap the smallest value in the unsorted part (2) with the first value in the unsorted part (6)

Swaps 2 and 6 1 6 5 2 4 3

3 is the smallest value in the unsorted part. Swap it with 5, the first value in the unsorted part

Swaps 3 and 5 1 2 5 6 4 3

5 isn't in the right position yet, but 4 is smaller so it's swapped first

Swaps 4 and 6 1 2 3 6 4 5

Swaps 5 and 6 1 2 3 4 6 5

The largest number is in last position after the swap, so no further swapping is required

Sorted! 1 2 3 4 5 6

EXPERT TIPS

Sorting in Python

There are lots of different sorting algorithms, each with different strengths and weaknesses. Python's "sort()" function uses an algorithm called "Timsort", named after its designer, Tim Peters. It's based on two sorting algorithms: "Insertion sort" and "Merge sort". Type in this code to see how it works.

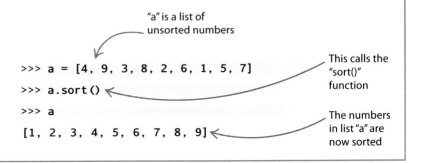

"a" is a list of unsorted numbers

```
>>> a = [4, 9, 3, 8, 2, 6, 1, 5, 7]
>>> a.sort()
>>> a
[1, 2, 3, 4, 5, 6, 7, 8, 9]
```

This calls the "sort()" function

The numbers in list "a" are now sorted

Libraries

Writing new code takes time, so it's useful to be able to reuse bits of other programs. These snippets of code can be shared in packages called "libraries".

SEE ALSO

Making **154–155 〉**
windows

Colour and **156–157 〉**
co-ordinates

Standard Library modules

Python comes with a "Standard Library" that has lots of useful bits of code ready to use. Stand-alone sections of a library called "modules" can be added to Python to make it even more powerful.

◁ **Batteries included**
Python's motto is "batteries are included". This means it comes with lots of ready-to-use code.

◁ **Random**
This module can pick a random number, or shuffle a list into a random order.

▽ **Turtle**
This module is used to draw lines and shapes on the screen.

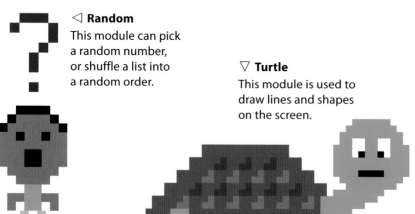

△ **Time**
The Time module gives the current time and date, and can calculate dates – for instance, what day will it be in three days' time?

▽ **Tkinter**
Tkinter is used to make buttons, windows, and other graphics that help users interact with programs.

▷ **Math**
Use the Math module to work with complex mathematical calculations.

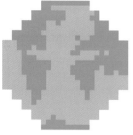

△ **Socket**
The code in this module helps computers connect to each other over networks and the internet.

Importing modules

Before using a module, you have to tell the computer to import it so it can be used by your program. This allows the bits of code it contains to be available to you. Importing modules is done using the "import" command. Python can import modules in a few different ways.

Pygame

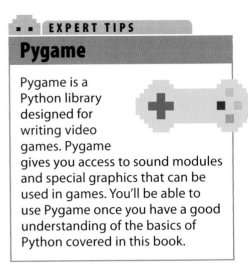

Pygame is a Python library designed for writing video games. Pygame gives you access to sound modules and special graphics that can be used in games. You'll be able to use Pygame once you have a good understanding of the basics of Python covered in this book.

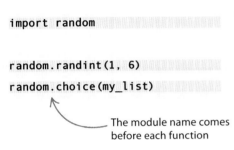

```
import random

random.randint(1, 6)
random.choice(my_list)
```

The module name comes before each function

◁ **"import random"**
This way of importing requires you to type the module name at the start of the code. It makes it easier to read because you know which module it came from.

Imports all the functions from the Random module

```
from random import *

randint(1, 6)
choice(my_list)
```

This code doesn't show which module the function came from

▷ **"from random import *"**
Importing a module like this works well for small programs. But it can get confusing with bigger programs, as it isn't clear which module the function belongs to.

Imports only the "randint" function

```
from random import randint

randint(1, 6)
```

Only the "randint" function is available

◁ **"from random import randint"**
You can import a single function from the module. This can be more efficient than importing the whole module if it's the only function you want to use.

Help and documentation

Not sure how to use a module or what functions are available? The Python Library Reference has all the details. Simply click on the library you want to learn more about. It's a good idea to get to know the libraries, modules, and functions that are available, so you don't waste time writing code that already exists.

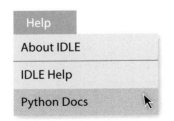

Help
About IDLE
IDLE Help
Python Docs

◁ **Help!**
At the top of any IDLE window, click "Help" and choose "Python Docs". This brings up a window with lots of useful information.

Making windows

Many programs have windows and buttons that can be used to control them. These make up the "graphical user interface", or "GUI" (pronounced "gooey").

SEE ALSO

Colour and **156–157)**
co-ordinates

Making **158–159)**
shapes

Changing **160–161)**
things

Make a simple window

The first step in creating a GUI is to make the window that will hold everything else inside it. Tkinter (from Python's Standard Library) can be used to create a simple one.

1 Enter the code
This code imports Tkinter from the library and creates a new window. Tkinter must be imported before it can be used.

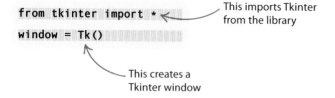

```
from tkinter import *
window = Tk()
```

This imports Tkinter from the library

This creates a Tkinter window

2 A Tkinter window appears
Run the code and a window appears. It looks a bit dull for now, but this is only the first part of your GUI.

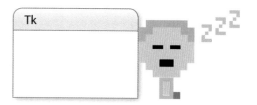

Add buttons to the window

Make the GUI more interactive by adding buttons. A different message will be displayed when the user clicks each button.

1 Create two buttons
Write this code to create a simple window with two buttons.

This message appears when button A is pressed

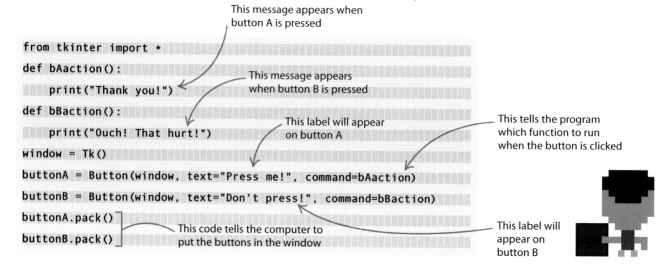

```
from tkinter import *
def bAaction():
    print("Thank you!")
def bBaction():
    print("Ouch! That hurt!")
window = Tk()
buttonA = Button(window, text="Press me!", command=bAaction)
buttonB = Button(window, text="Don't press!", command=bBaction)
buttonA.pack()
buttonB.pack()
```

This message appears when button B is pressed

This label will appear on button A

This tells the program which function to run when the button is clicked

This code tells the computer to put the buttons in the window

This label will appear on button B

2 **Click the buttons**
to print messages

When the program is run, a window with two buttons appears. Click the buttons and different messages will appear in the shell. You've now made an interactive GUI that responds to the user's commands.

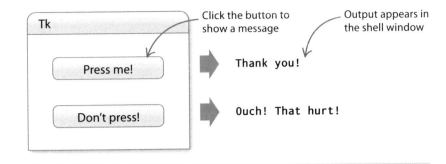

Click the button to show a message

Output appears in the shell window

Tk

Press me!

Thank you!

Don't press!

Ouch! That hurt!

Roll the dice

Tkinter can be used to build a GUI for a simple application. The code below creates a program that simulates rolling a six-sided dice.

1 **Create a dice simulator**
This program creates a button that, when pressed, tells the function "roll()" to display a random number between 1 and 6.

This imports the function "randint" from the random library

```
from tkinter import *
from random import randint
def roll():
    text.delete(0.0, END)
    text.insert(END, str(randint(1,6)))
window = Tk()
text = Text(window, width=1, height=1)
buttonA = Button(window, text="Press to roll!", command=roll)
text.pack()
buttonA.pack()
```

This code clears the text inside the text box and replaces it with a random number between 1 and 6

Creates a text box to display the random number

This tells the program which function to run when the button is clicked

This puts the text box and the button in the window

This label appears on the button

2 **Press the button to**
roll the dice

Run the program, then click the button to roll the dice and see the result. This program can be simply changed so that it simulates a 12-sided dice, or a coin being tossed.

A new number appears here each time the button is clicked

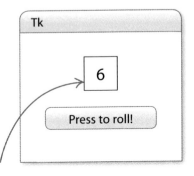

Tk

6

Press to roll!

■ ■ **EXPERT TIPS**

Clear and simple

When you're designing a GUI, try not to confuse the user by filling the screen with too many buttons. Label each button with a sensible name to make the application easy to understand.

Colour and co-ordinates

Pictures and graphics on a computer screen are made up of tiny coloured dots called pixels. To create graphics in a program, the computer needs to be told exactly what colour each pixel should be.

SEE ALSO

❮ **154–155** Making windows

Making **158–159** ❯ shapes

Changing **160–161** ❯ things

Selecting colours

It's important to describe colours in a way that computers can understand. Tkinter includes a useful tool to help you do this.

1 Launch the colour selection tool
Type the following code into the shell window to launch the Tkinter tool for selecting colours.

This imports the colour selecting function of Tkinter

```
>>> import tkinter.colorchooser
>>> tkinter.colorchooser.askcolor()
```

Use the American spelling of colour

2 Choose a colour
The "color chooser" window will appear. Pick the colour you want and then click the "OK" button.

Select the colour you want by clicking on it

This window makes it easy to pick the exact colour you want

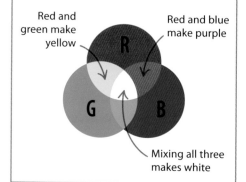

•• **EXPERT TIPS**

Mixing colours

Each pixel can give out red, green, and blue light. By mixing these colours together, you can make any colour imaginable.

Red and green make yellow

Red and blue make purple

Mixing all three makes white

3 Colour values
When a colour is selected, a list of numbers will appear in the shell window. These numbers are the values of red, green, and blue that have been mixed to make the chosen colour.

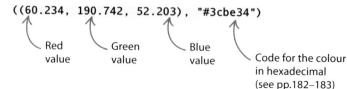

```
((60.234, 190.742, 52.203), "#3cbe34")
```

Red value

Green value

Blue value

Code for the colour in hexadecimal (see pp.182–183)

Drawing on a canvas

To create graphics using Python, you need to make a blank area to draw on. This is known as a canvas. You can use x and y co-ordinates to tell Python exactly where to draw on the canvas.

· · EXPERT TIPS

Co-ordinates

In Tkinter, x co-ordinates get larger moving to the right, and y co-ordinates get larger moving downwards. (0,0) is in the top-left corner.

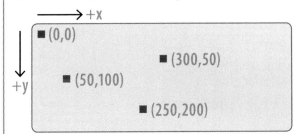

1 Create a graphics program
Use this code to create a window and put a canvas inside it. It will then draw random circles on the canvas.

This imports the "randint" and "choice" functions from the Random module

```
from random import *
from tkinter import *
size = 500
window = Tk()
canvas = Canvas(window, width=size, height=size)
canvas.pack()
while True:
    col = choice(["pink", "orange", "purple", "yellow"])
    x0 = randint(0, size)
    y0 = randint(0, size)
    d = randint(0, size/5)
    canvas.create_oval(x0, y0, x0 + d, y0 + d, fill=col)
    window.update()
```

This imports all of the Tkinter functions

The variable "size" sets the dimensions of the canvas

This creates a canvas inside a window

A forever loop makes the program draw circles endlessly

This chooses a random colour from the list

This creates a circle of a random size in a random place on the canvas

This part fills it with the colour that has been chosen ("col")

This part of the line draws the circle

2 Coloured canvas
Run the code and the program will start drawing circles on the canvas.

Tk

The size of each circle is random

Circles are drawn in random places

Making shapes

As well as adding windows, buttons, and colours to a graphical user interface (GUI), Tkinter can also be used to draw shapes.

SEE ALSO

Changing **160–161 ⟩**
things

Reacting **162–163 ⟩**
to events

Creating basic shapes

Rectangles and ovals are useful shapes for drawing all sorts of things. Once a canvas has been created, the following functions can be used to draw shapes on it.

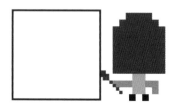

```
>>> from tkinter import *
>>> window = Tk()
>>> drawing = Canvas(window, height=500, width=500)
>>> drawing.pack()
>>> rect1 = drawing.create_rectangle(100, 100, 300, 200)
>>> square1 = drawing.create_rectangle(30, 30, 80, 80)
>>> oval1 = drawing.create_oval(100, 100, 300, 200)
>>> circle1 = drawing.create_oval(30, 30, 80, 80)
```

Creates a canvas to draw on

Sets the size of the canvas

Draws a rectangle

Sets the position and size of the rectangle using co-ordinates (see below)

A square can be made by drawing a rectangle with all sides the same length

Draws a circle

Sets the position and size of the circle

Drawing with co-ordinates

Co-ordinates are used to tell the computer exactly where to create shapes. The first number ("x") tells the computer how far along the screen to go. The second number ("y") tells the computer how far down to go.

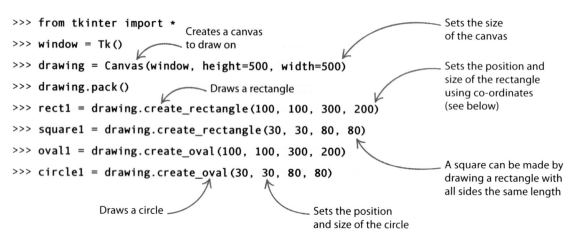

This is the name of the canvas

Co-ordinates for the top left of the rectangle

```
>>> drawing.create_rectangle(50, 50, 250, 350)
```

△ **Setting the co-ordinates**
The first two numbers give the co-ordinates for the top-left corner of the rectangle. The second two numbers locate the bottom-right corner.

Co-ordinates for the bottom right of the rectangle

▽ **Co-ordinates grid**
The top-left corner of the rectangle is at co-ordinates (50, 50). The bottom-right corner is at (350, 250).

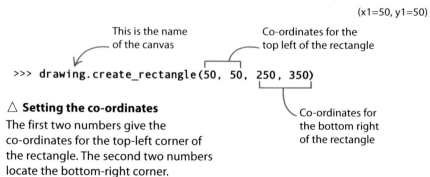

(x1=50, y1=50)

(x2=350, y2=250)

Adding colour to shapes

It's also possible to create coloured shapes. Code can be used to set different colours for the outline and the inside ("fill") of each shape.

Creates a solid blue circle with a red outline

```
>>> drawing.create_oval(30, 30, 80, 80, outline="red", fill="blue")
```

Draw an alien

You can draw almost anything by combining different shapes. Here are some instructions for creating an alien using ovals, lines, and triangles.

1 Create the alien
For each part of the alien, you must define the type of shape, size, position on the canvas, and colour. Each shape has a unique ID number that can be stored in a variable.

Creates the canvas

Sets "Alien" as the title of the window

```
from tkinter import *
window = Tk()
window.title("Alien")
c = Canvas(window, height=300, width=400)
c.pack()
body = c.create_oval(100, 150, 300, 250, fill="green")
eye = c.create_oval(170, 70, 230, 130, fill="white")
eyeball = c.create_oval(190, 90, 210, 110, fill="black")
mouth = c.create_oval(150, 220, 250, 240, fill="red")
neck = c.create_line(200, 150, 200, 130)
hat = c.create_polygon(180, 75, 220, 75, 200, 20, fill="blue")
```

Draws a green oval for the body

Draws a black dot inside the eye

Draws a red oval for the mouth

Draws a blue triangle for the alien's hat

2 Meet the alien
Run the code to draw the alien. It has a green body, a red mouth, and one eye on a stalk. It's also wearing a lovely blue hat.

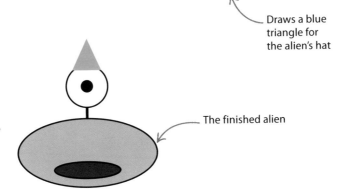

The finished alien

Changing things

Once a graphic has been drawn on the canvas, it doesn't need to stay the same. Code can be used to change the way it looks, or move it around the screen.

SEE ALSO

⟨ **158–159** Making shapes

Reacting to **162–163** ⟩ events

Moving shapes

To make a shape move on the canvas, you need to tell the computer what to move (the name or ID you gave the shape) and where to move it.

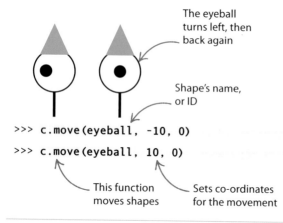

The eyeball turns left, then back again

Shape's name, or ID

```
>>> c.move(eyeball, -10, 0)
>>> c.move(eyeball, 10, 0)
```

This function moves shapes

Sets co-ordinates for the movement

REMEMBER

Meaningful names

It's a good idea to use sensible names to identify the shapes on the canvas. These pages use names like "eyeball" and "mouth" so the code is easy to read and understand.

◁ **Moving eyeballs**
Type this code into the shell window to make the eyeball turn to the left, then turn back again.

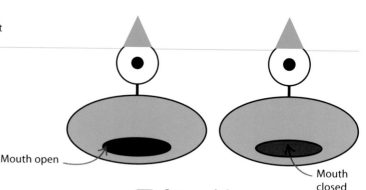

Mouth open

Mouth closed

Changing colours

You can make the mouth look as though it is opening and closing by simply changing the colour of the oval.

1 **Write the code**
Type this code to create two functions that will make the mouth seem to open and close.

The function "itemconfig()" changes the properties of shapes you've already drawn

```
def mouth_open():
    c.itemconfig(mouth, fill="black")
def mouth_close():
    c.itemconfig(mouth, fill="red")
```

The opened mouth will be black

The closed mouth will be red

The shape's ID

2 **Open and close**
Type this code into the shell window to make the mouth open and close.

```
>>> mouth_open()
>>> mouth_close()
```

Enter these commands to make the alien open and close its mouth

Hide and show

Shapes can be hidden using the "itemconfig()" function. If you hide the eyeball, and then show it again a moment later, the alien looks as though it is blinking.

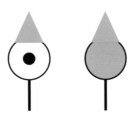

◁ **Blinking alien**
To make the alien blink, you need to hide the pupil and make the white of the eye green.

1 **Create blinking functions**
This code creates two functions so you can make the alien blink.

The shape's ID

```
def blink():
    c.itemconfig(eye, fill="green")
    c.itemconfig(eyeball, state=HIDDEN)
def unblink():
    c.itemconfig(eye, fill="white")
    c.itemconfig(eyeball, state=NORMAL)
```

Turns the white of the eye green

Hides the pupil

Makes the eye white again

Reveals the pupil

2 **Blink and unblink**
Type this code into the shell window to make the alien blink.

```
>>> blink()
>>> unblink()
```

The "unblink()" command makes the eye appear open again

Saying things

Text can also be displayed on the screen to make the alien talk. You can even make it say different things in response to user commands.

I am an alien!

1 **Adding text**
This code adds text to the graphic of the alien and creates a function to steal its hat.

Positions the text on the canvas

```
words = c.create_text(200, 280, text="I am an alien!")
def steal_hat():
    c.itemconfig(hat, state=HIDDEN)
    c.itemconfig(words, text="Give my hat back!")
```

This hides the hat

Put what you want the alien to say in quote marks

As soon as the hat disappears the alien will ask for it back

A new message appears when the hat disappears

Give my hat back!

2 **Steal the hat**
Type this code into the shell window and see what happens.

Type this to steal the hat

```
>>> steal_hat()
```

Reacting to events

Computers receive a signal when a key is pressed or a mouse is moved. This is called an "event". Programs can instruct the computer to respond to any events it detects.

SEE ALSO

❮ **158–159** Making shapes

❮ **160–161** Changing things

Event names

Lots of different events can be triggered using input devices like a mouse or keyboard. Tkinter has names to describe each of these events.

Mouse events

\<Button-1\>
↖ Left mouse button clicked

\<Button-3\>
↖ Right mouse button clicked

Keyboard events

\<Right\>
Right arrow key pressed ↗

\<Left\>
Left arrow key pressed ↗

Spacebar pressed ↘ **\<space\>**

\<Up\>
Up arrow key pressed ↗

Down arrow key pressed ↓
\<Down\>

"A" key pressed ↘   Different letters can go here ↘
\<KeyPress-a\>

Mouse events

To make a program respond to mouse events, simply link (or bind) a function to an event. Here, the function "burp" is created, then bound to the "\<Button-1\>" event.

This brings the Tkinter window to the front of your screen ↘

Creates a function called "burp" ↘

```
window.attributes("-topmost", 1)
def burp(event):
    mouth_open()
    c.itemconfig(words, text="Burp!")
c.bind_all("<Button-1>", burp)
```

Links the left mouse click to the "burp" function ↘

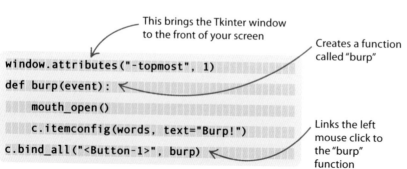

△ **Burping alien**
Click the left mouse button and the alien lets out a burp. This is because the "burp" function has been used.

Key events

Functions can also be bound to keys on the keyboard in the same way. Type in the code below to make the alien blink when the "A" and "Z" keys are pressed.

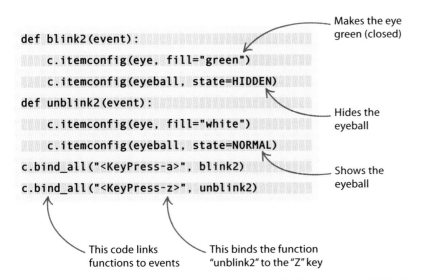

```
def blink2(event):
    c.itemconfig(eye, fill="green")
    c.itemconfig(eyeball, state=HIDDEN)
def unblink2(event):
    c.itemconfig(eye, fill="white")
    c.itemconfig(eyeball, state=NORMAL)
c.bind_all("<KeyPress-a>", blink2)
c.bind_all("<KeyPress-z>", unblink2)
```

Makes the eye green (closed)

Hides the eyeball

Shows the eyeball

This code links functions to events

This binds the function "unblink2" to the "Z" key

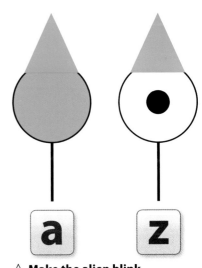

△ **Make the alien blink**
When this code is run, the "A" key will make the eye close, and the "Z" key will make it open again.

Moving with keys

Key presses can also be used to trigger movement. This code binds the arrow keys to functions that make the alien's eyeball move.

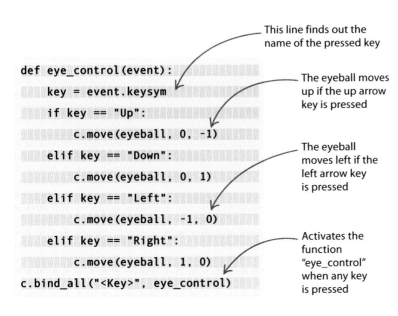

```
def eye_control(event):
    key = event.keysym
    if key == "Up":
        c.move(eyeball, 0, -1)
    elif key == "Down":
        c.move(eyeball, 0, 1)
    elif key == "Left":
        c.move(eyeball, -1, 0)
    elif key == "Right":
        c.move(eyeball, 1, 0)
c.bind_all("<Key>", eye_control)
```

This line finds out the name of the pressed key

The eyeball moves up if the up arrow key is pressed

The eyeball moves left if the left arrow key is pressed

Activates the function "eye_control" when any key is pressed

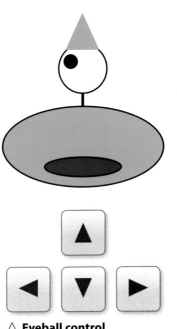

△ **Eyeball control**
The eyeball moves in the direction of the pressed arrow key.

SEE ALSO

❮ **154–155** Making windows

❮ **156–157** Colour and co-ordinates

❮ **158–159** Making shapes

PROJECT 7

Bubble blaster

This project uses all the skills taught in this chapter to make a game. It's a big project, so tackle it in stages and remember to save the program regularly. Try to understand how each part fits together before moving on to the next stage. By the end you'll have a game that you can play and share with friends.

Aim of the game

Before writing any code, think about the overall plan for the game and how it should work. Here are the main rules that set out how the game will be played:

The player controls a submarine

The arrow keys move the submarine

Popping bubbles scores points

A timer is set to 30 seconds at the start

Scoring 1,000 points earns extra time

The game ends when the time runs out

Create the game window and the submarine

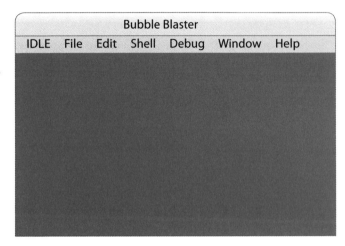

Bubble Blaster

IDLE File Edit Shell Debug Window Help

Start by setting the scene. Open a new code window in IDLE. Type in the code below to create the window for the game, and the submarine that the player controls.

1 Use the Tkinter library to build the graphical user interface (GUI). This code will create the main window for the game.

```
from tkinter import *
HEIGHT = 500
WIDTH = 800
window = Tk()
window.title("Bubble Blaster")
c = Canvas(window, width=WIDTH, height=HEIGHT, bg="darkblue")
c.pack()
```

Imports all of the Tkinter functions

Sets the size of the window

Give the game a snappy title

Sets dark blue as the colour of the background (the sea)

Creates a canvas that can be drawn on

The submarine will be represented by a triangle inside a circle

2 A simple graphic will represent the submarine in this game. This can be made using some of the drawing functions from Tkinter. Type out this code, then run it.

Draws a red triangle for the submarine

Draws a red circle outline

```
ship_id = c.create_polygon(5, 5, 5, 25, 30, 15, fill="red")
ship_id2 = c.create_oval(0, 0, 30, 30, outline="red")
SHIP_R = 15
MID_X = WIDTH / 2
MID_Y = HEIGHT / 2
c.move(ship_id, MID_X, MID_Y)
c.move(ship_id2, MID_X, MID_Y)
```

The radius (size) of the submarine

The variables "MID_X" and "MID_Y" give the co-ordinates of the middle of the screen

Moves both parts of the submarine to the centre of the screen

Don't forget to save your work

⊳ BUBBLE BLASTER

Controlling the submarine

The next stage of the program is to write the code that makes the submarine move when the arrow keys are pressed. The code will create a function called an "event handler". The event handler checks which key has been pressed and moves the submarine.

3 Type this code to create a function called "move_ship". This function will move the submarine in the correct direction when a cursor key is pressed. Try running it to see how it works.

The sub will move this far when a key is pressed

```
SHIP_SPD = 10

def move_ship(event):
    if event.keysym == "Up":
        c.move(ship_id, 0, -SHIP_SPD)
        c.move(ship_id2, 0, -SHIP_SPD)
    elif event.keysym == "Down":
        c.move(ship_id, 0, SHIP_SPD)
        c.move(ship_id2, 0, SHIP_SPD)
    elif event.keysym == "Left":
        c.move(ship_id, -SHIP_SPD, 0)
        c.move(ship_id2, -SHIP_SPD, 0)
    elif event.keysym == "Right":
        c.move(ship_id, SHIP_SPD, 0)
        c.move(ship_id2, SHIP_SPD, 0)
c.bind_all("<Key>", move_ship)
```

Moves the two parts of the sub up when the up arrow key is pressed

These lines are activated when the down arrow key is pressed, and the sub moves down

The sub moves left when the left arrow key is pressed

Moves the sub right when the right arrow key is pressed

Tells Python to run "move_ship" whenever any key is pressed

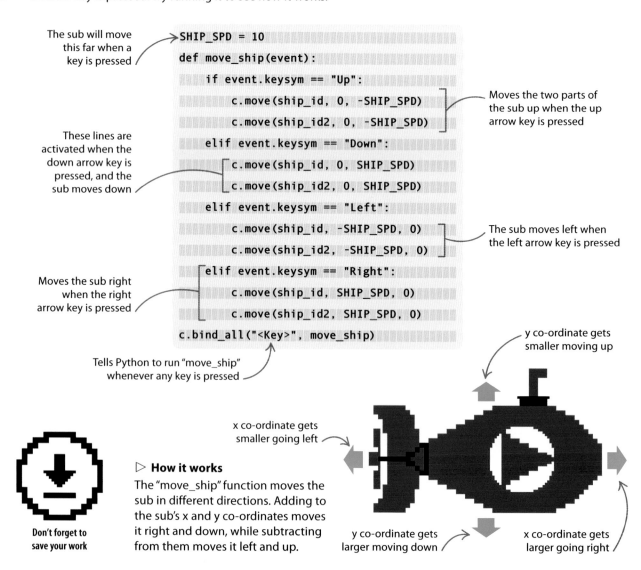

y co-ordinate gets smaller moving up

x co-ordinate gets smaller going left

y co-ordinate gets larger moving down

x co-ordinate gets larger going right

Don't forget to save your work

▷ **How it works**

The "move_ship" function moves the sub in different directions. Adding to the sub's x and y co-ordinates moves it right and down, while subtracting from them moves it left and up.

Get ready for bubbles

Now the submarine can move, start creating the bubbles for the player to pop. Each bubble will be a different size and move at a different speed.

 4 Every bubble needs an ID number (so the program can identify each specific bubble), a size, and a speed.

This creates three empty lists used to store the ID, radius (size), and speed of each bubble

Sets the minimum radius of the bubble to 10, and the maximum to 30

Picks a random size for the bubble, between the maximum and minimum values possible

Sets the position of the bubble on the canvas

This line of code creates the bubble shape

```
from random import randint
bub_id = list()
bub_r = list()
bub_speed = list()
MIN_BUB_R = 10
MAX_BUB_R = 30
MAX_BUB_SPD = 10
GAP = 100
def create_bubble():
    x = WIDTH + GAP
    y = randint(0, HEIGHT)
    r = randint(MIN_BUB_R, MAX_BUB_R)
    id1 = c.create_oval(x - r, y - r, x + r, y + r, outline="white")
    bub_id.append(id1)
    bub_r.append(r)
    bub_speed.append(randint(1, MAX_BUB_SPD))
```

Adds the ID, radius, and speed of the bubble to the three lists

.. ▪ EXPERT TIPS

Bubble lists

Three lists are used to store information about each bubble. The lists start off empty, and information about each bubble is then added as you create it. Each list stores a different bit of information.

bub_id: stores the ID number of the bubble so the program can move it later.

bub_r: stores the radius (size) of the bubble.

bub_speed: stores how fast the bubble travels across the screen.

Don't forget to save your work

BUBBLE BLASTER

Make the bubbles move

There are now lists to store the ID, size, and speed of the bubbles, which are randomly generated. The next stage is to write the code that makes the bubbles move across the screen.

Goes through each bubble in the list

5 This function will go through the list of bubbles and move each one in turn.

```python
def move_bubbles():
    for i in range(len(bub_id)):
        c.move(bub_id[i], -bub_speed[i], 0)
```

Moves the bubble across the screen according to its speed

Imports the functions you need from the Time library

6 This will be the main loop for the game. It will be repeated over and over while the game is running. Try running it!

```python
from time import sleep, time
BUB_CHANCE = 10
#MAIN GAME LOOP
while True:
    if randint(1, BUB_CHANCE) == 1:
        create_bubble()
    move_bubbles()
    window.update()
    sleep(0.01)
```

Generates a random number from 1 to 10

If the random number is 1, the program creates a new bubble (on average 1 in 10 times – so there aren't too many bubbles!)

Runs the "move_bubbles" function

Don't forget to save your work

Slows the game down so it's not too fast to play

Updates the window to redraw bubbles that have moved

7 Now you're going to create a useful function to find out where a particular bubble is, based on the ID. This code should be added to the program directly after the code you created in step 5.

```python
def get_coords(id_num):
    pos = c.coords(id_num)
    x = (pos[0] + pos[2])/2
    y = (pos[1] + pos[3])/2
    return x, y
```

Works out the x co-ordinate of the middle of the bubble

Works out the y co-ordinate of the middle of the bubble

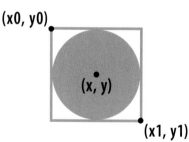

△ **Locating bubbles**
The function finds the middle of the bubble by taking the point halfway between the corners of the box around it.

How to make bubbles pop

The player will score points when the bubbles are popped, so the program has to make bubbles disappear from the screen. These next functions will allow it to do that.

8 This function will be used to remove a bubble from the game. It does this by deleting it from all the lists, and from the canvas. This code should be added directly after the code you typed out in step 7.

This function deletes the bubble with ID "i"

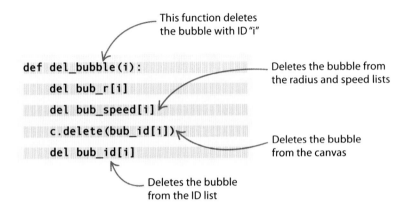

```
def del_bubble(i):
    del bub_r[i]
    del bub_speed[i]
    c.delete(bub_id[i])
    del bub_id[i]
```

Deletes the bubble from the radius and speed lists

Deletes the bubble from the canvas

Deletes the bubble from the ID list

9 Type this code to create a function that cleans up bubbles that have floated off the screen. This code should go directly after the code from step 8.

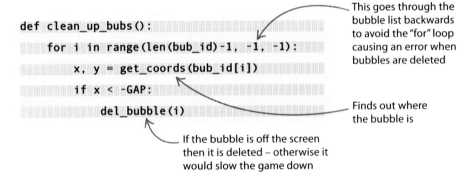

```
def clean_up_bubs():
    for i in range(len(bub_id)-1, -1, -1):
        x, y = get_coords(bub_id[i])
        if x < -GAP:
            del_bubble(i)
```

This goes through the bubble list backwards to avoid the "for" loop causing an error when bubbles are deleted

Finds out where the bubble is

If the bubble is off the screen then it is deleted – otherwise it would slow the game down

10 Now update the main game loop (from step 6) to include the helpful functions you have just created. Run it to make sure you haven't included any errors.

Removes bubbles that are off the screen

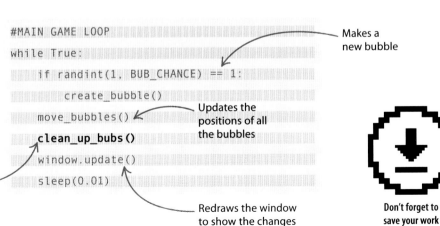

```
#MAIN GAME LOOP
while True:
    if randint(1, BUB_CHANCE) == 1:
        create_bubble()
    move_bubbles()
    clean_up_bubs()
    window.update()
    sleep(0.01)
```

Makes a new bubble

Updates the positions of all the bubbles

Redraws the window to show the changes

Don't forget to save your work

⊙ BUBBLE BLASTER

Working out the distance between points

In this game, and lots of others, it is useful to know the distance between two objects. Here's how to use a well-known mathematical formula to have the computer work it out.

11 This function calculates the distance between two objects. Add this bit of code directly after the code you wrote in step 9.

Loads the "sqrt" function from the Math library

Gets the position of the first object

Gets the position of the second object

```
from math import sqrt
def distance(id1, id2):
    x1, y1 = get_coords(id1)
    x2, y2 = get_coords(id2)
    return sqrt((x2 - x1)**2 + (y2 - y1)**2)
```

Gives back the distance between them

Pop the bubbles

The player scores points by popping bubbles. Big bubbles and fast bubbles are worth more points. The next section of code works out when each bubble is popped by using its radius (the distance from the centre to the edge).

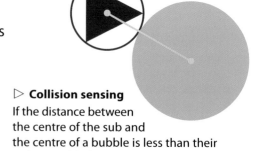

▷ **Collision sensing**
If the distance between the centre of the sub and the centre of a bubble is less than their radiuses added together, they have collided.

12 When the submarine and a bubble crash into each other, the program needs to pop the bubble and update the score. This bit of code should come directly after the code in step 11.

This variable keeps track of points scored

This loop goes through the entire list of bubbles (it goes backwards to avoid errors when deleting bubbles)

Checks for collisions between the sub and any bubbles

```
def collision():
    points = 0
    for bub in range(len(bub_id)-1, -1, -1):
        if distance(ship_id2, bub_id[bub]) < (SHIP_R + bub_r[bub]):
            points += (bub_r[bub] + bub_speed[bub])
            del_bubble(bub)
    return points
```

Gives back the number of points

Deletes the bubble

Calculates the number of points this bubble is worth and adds it to "points"

13 Now update the main game loop to use the functions you have just created. Remember that the order is important, so make sure you put everything in the right place. Then run the code. Bubbles should burst when they hit the sub. Check the shell window to see the score.

Sets the score to zero when the game starts

```
score = 0
#MAIN GAME LOOP
while True:
    if randint(1, BUB_CHANCE) == 1:
        create_bubble()
    move_bubbles()
    clean_up_bubs()
    score += collision()
    print(score)
    window.update()
    sleep(0.01)
```

Creates new bubbles

Adds the bubble score to the total

Shows the score in the shell window – it will be displayed properly later

This pauses the action for a very short time – try removing this and see what happens

Don't forget to save your work

BUBBLE BLASTER

Adding a few final touches

The main stages of the game are now working. All that remains is to add the final parts: displaying the player's score, and setting a time limit that counts down until the game ends.

14 ▸ Type in this code after the code you entered in step 12. It tells the computer to display the player's score and the time left in the game.

Creates "TIME" and "SCORE" labels to explain to the player what the numbers mean

```
c.create_text(50, 30, text="TIME", fill="white" )
c.create_text(150, 30, text="SCORE", fill="white" )
time_text = c.create_text(50, 50, fill="white" )
score_text = c.create_text(150, 50, fill="white" )
def show_score(score):
    c.itemconfig(score_text, text=str(score))
def show_time(time_left):
    c.itemconfig(time_text, text=str(time_left))
```

Sets the scores and time remaining

Displays the score

Displays the time remaining

15 ▸ Next, set up the time limit and the score required to gain bonus time, and calculate the end time of the game. This bit of code should come just before the main game loop.

Imports functions from the Time library

```
from time import sleep, time
BUB_CHANCE = 10
TIME_LIMIT = 30
BONUS_SCORE = 1000
score = 0
bonus = 0
end = time() + TIME_LIMIT
```

Starts the game with a 30-second time limit

Sets when bonus time is given (when a player has scored 1,000 points)

Stores the finish time in a variable called "end"

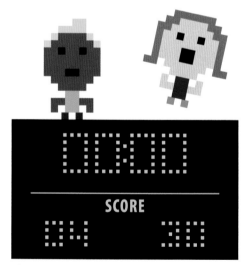

△ **Scoreboard**
Scoreboards are a great visual way to show the player at a glance how well they are doing in a game.

16 Update the main game loop to include the new score and time functions.

```
#MAIN GAME LOOP
while time() < end:
    if randint(1, BUB_CHANCE) == 1:
        create_bubble()
    move_bubbles()
    clean_up_bubs()
    score += collision()
    if (int(score / BONUS_SCORE)) > bonus:
        bonus += 1
        end += TIME_LIMIT
    show_score(score)
    show_time(int(end - time()))
    window.update()
    sleep(0.01)
```

Repeats the main game loop until the game ends

Calculates when to give bonus time

"print(score)" has been replaced by "show_score(score)" so that the score now appears in the game window

Displays the time remaining

Don't forget to save your work

17 Finally, add a "GAME OVER" graphic. This will be shown when the time runs out. Add this to the very bottom of your program.

Puts graphic in the middle of the screen

```
c.create_text(MID_X, MID_Y, \
    text="GAME OVER", fill="white", font=("Helvetica",30))
c.create_text(MID_X, MID_Y + 30, \
    text="Score: "+ str(score), fill="white")
c.create_text(MID_X, MID_Y + 45, \
    text="Bonus time: "+ str(bonus*TIME_LIMIT), fill="white")
```

Sets the font – "Helvetica" is a good font for big letters

Tells you what your score was

Shows how much bonus time was earned

Sets the text colour to white

Don't forget to save your work

❯ BUBBLE BLASTER

Time to play

Well done! You've finished writing Bubble blaster and it's now ready to play. Run the program and try it out. If something isn't working, remember the debugging tips – look back carefully over the code on the previous pages to make sure everything is typed out correctly.

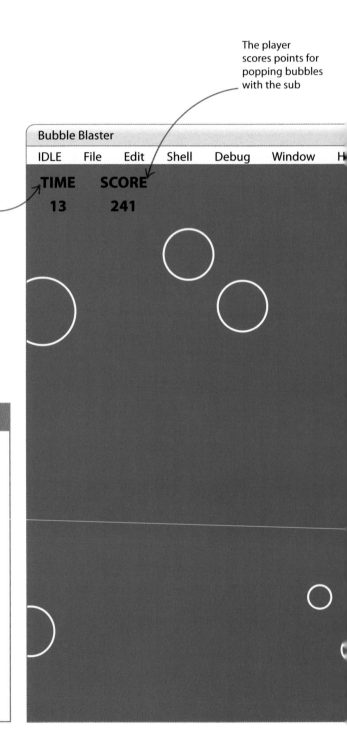

The player scores points for popping bubbles with the sub

The timer counts down to the end of the game

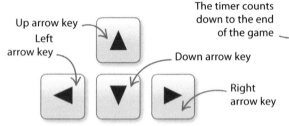

Up arrow key

Left arrow key

Down arrow key

Right arrow key

△ **Controls**

The submarine is steered using the arrow keys. The program can be adjusted so it works with other controls.

▪ ▪ EXPERT TIPS

Improving your game

All computer games start as a basic idea. They are then played, tested, adjusted, and improved. Think of this as version one of your game. Here are some suggestions of how you could change and improve it with new code:

Make the game harder by adjusting the time limit and the score required for bonus time.

Choose a different colour for your submarine.

Create a more detailed submarine graphic.

Have a special type of bubble that increases the speed of the submarine.

Add a smart bomb that deletes all of the bubbles when you press the spacebar.

Build a leaderboard to keep track of the best scores.

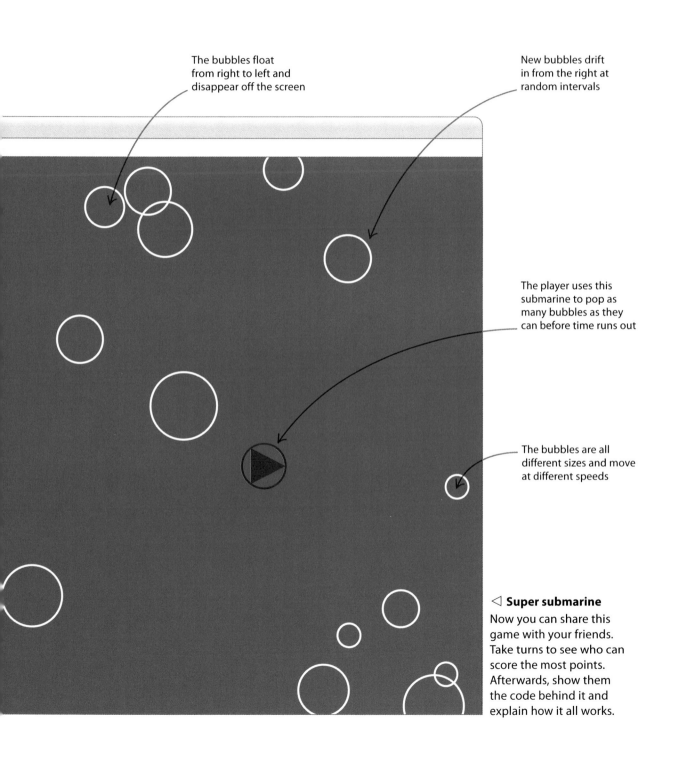

The bubbles float from right to left and disappear off the screen

New bubbles drift in from the right at random intervals

The player uses this submarine to pop as many bubbles as they can before time runs out

The bubbles are all different sizes and move at different speeds

◁ **Super submarine**
Now you can share this game with your friends. Take turns to see who can score the most points. Afterwards, show them the code behind it and explain how it all works.

What next?

Now that you've tackled the Python projects in this book, you're on your way to becoming a great programmer. Here are some ideas for what to do next in Python, and how to take your programming skills further.

SEE ALSO

❬ **152–153** Libraries

Computer **204–205** ❭
games

Experiment

Play around with the code samples in this book. Find new ways to remix them or add new features – and don't be afraid to break them too! This is your chance to experiment with Python. Remember that it is a professional programming language with a lot of power – you can do all sorts of things with it.

Build your own libraries

Programmers love to reuse code and share their work. Create your own library of useful functions and share it. It's a great feeling to see your code being used by another programmer. You might build something as useful as Tkinter or Turtle!

⠂⠄⠄ REMEMBER
Read lots of code

Find interesting programs or libraries written by other people and read through the code and their comments. Try to understand how the code works, and why it is built that way. This increases your knowledge of coding practices. You will also learn useful bits of information about libraries that you can use in future programs.

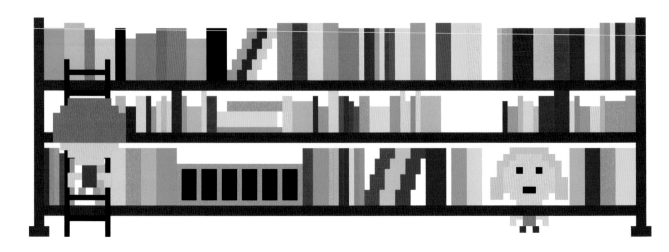

Make games with Python

You could create your own game using Python. The PyGame library, which is available to download from the web, comes with lots of functions and tools that make it easier to build games. Start by making simple games, then progress to more complex ones.

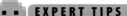 **EXPERT TIPS**

Different versions of Python

When you find code elsewhere (in other books or online), it may be written for a different version of Python. The versions are similar, but you might need to make small changes.

```
print "Hello World"
```
Python 2

```
print("Hello World")
```
Python 3

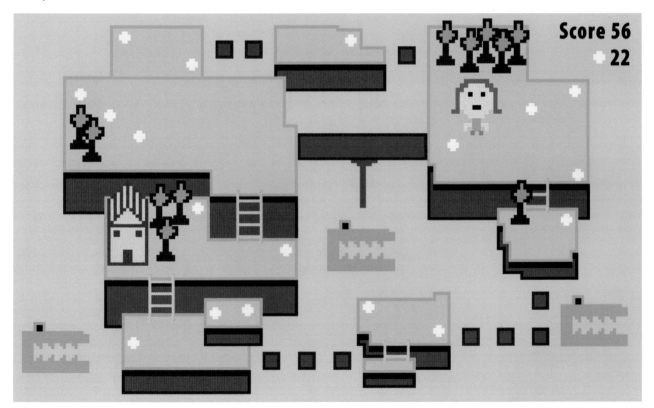

Score 56
22

Debug your code

Debugging is an important part of programming. Don't just give up if something isn't working. Remember that computers will only do what you tell them, so look through the code and figure out why it's not working. Sometimes looking over it with another programmer helps you to find bugs quicker.

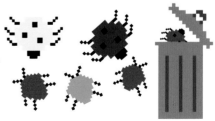

Inside computers

Inside a computer

The earliest computers were simple calculators. At a basic level, computers haven't changed much since then. They take in data (input), perform calculations, and give out answers (output).

SEE ALSO

Storing data **192–193 ⟩**
in files

The internet **194–195 ⟩**

Mini **214–215 ⟩**
computers

Basic elements

A computer consists of four main parts: input, memory, processor, and output. Input devices gather data, similar to the way your eyes or ears collect information about the world around you. Memory stores the data, while processors examine and alter it, just like a human brain. Output devices show the results of the processor's calculations, like a person speaking or moving after deciding what to do.

▷ **Von Neumann architecture**
A scientist called John von Neumann first came up with the standard layout for a computer in 1945. His plan is still followed today, with some improvements.

The memory contains information in sections, like books on library shelves. Memory is used to store programs and the data they use

Memory

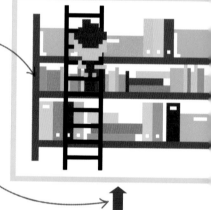

The control unit retrieves programs from the memory in order to run them

The control unit loads and carries out instructions from programs

Processor

Control unit

Input

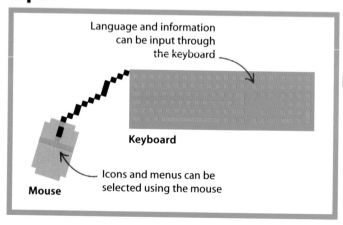

Language and information can be input through the keyboard

Keyboard

Icons and menus can be selected using the mouse

Mouse

Computer hardware

Hardware is the physical parts of a computer. Computers contain many different bits of hardware working together. As computer makers pack more and more features into smaller machines, the hardware components have to be smaller, generate less heat, and use less power.

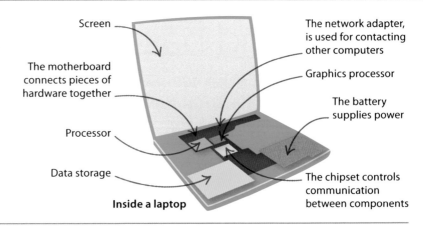

Screen

The network adapter, is used for contacting other computers

The motherboard connects pieces of hardware together

Graphics processor

The battery supplies power

Processor

Data storage

The chipset controls communication between components

Inside a laptop

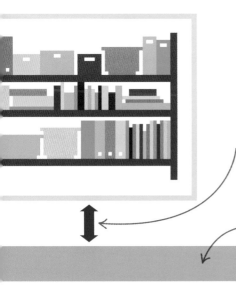

The arithmetic logic unit retrieves data for its calculations from the memory

The processor is made up of two parts, one to carry out instructions and the other to perform calculations

The arithmetic logic unit (ALU) performs any calculations the program needs

Arithmetic logic unit

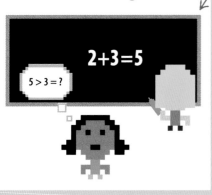

$2+3=5$

$5 > 3 = ?$

LINGO

GIGO

"Garbage in, garbage out" ("GIGO" for short) is a computing phrase meaning that even the best programs will output nonsense if they receive the wrong input.

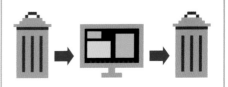

Output

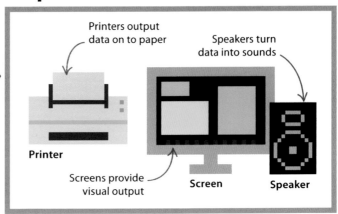

Printers output data on to paper

Speakers turn data into sounds

Printer

Screens provide visual output

Screen

Speaker

Binary and bases

How can computers solve complex calculations when all they understand is electrical signals? Binary numbers are used to translate these signals into numbers.

SEE ALSO

Symbols **184–185 ❭**
and codes

Logic gates **186–187 ❭**

What is a base number?

A "base" is the number of values that can be shown using only one digit. Each extra digit increases the number of values that can be shown by a multiple of the base.

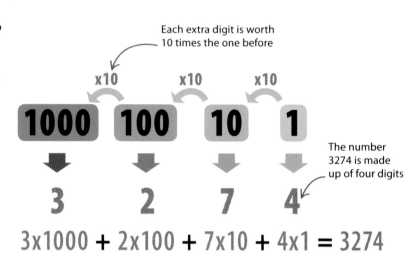

Each extra digit is worth 10 times the one before

x10   x10   x10

1000   100   10   1

The number 3274 is made up of four digits

▷ **Decimal system**
The decimal system is the most familiar counting system, and has a base of 10. It can show 10 values with one digit, 100 values with two digits, and 1000 with three digits.

3   2   7   4

3x1000 + 2x100 + 7x10 + 4x1 = 3274

Binary code

At the most basic level, computers only understand two values: electrical signals that are "on" and "off". As there are only two values, computers deal with numbers using a base of two, or "binary". Each digit is either a 1 or a 0, and each extra digit in the number is worth two times the previous digit.

A wire with a current

1
ON

▷ **1 and 0**
A wire with electrical signal "on" is a 1. A wire with electrical signal "off" is a 0.

0
OFF

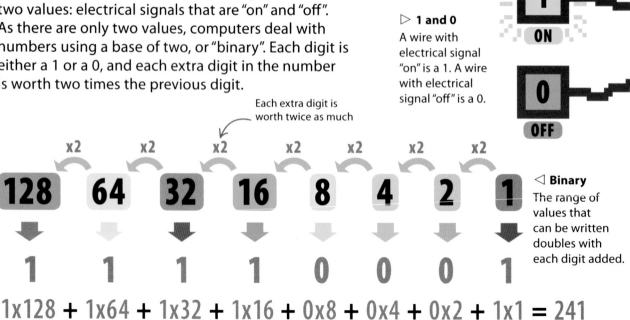

Each extra digit is worth twice as much

x2   x2   x2   x2   x2   x2   x2

128   64   32   16   8   4   2   1

◁ **Binary**
The range of values that can be written doubles with each digit added.

1   1   1   1   0   0   0   1

1x128 + 1x64 + 1x32 + 1x16 + 0x8 + 0x4 + 0x2 + 1x1 = 241

Hexadecimal

When using numbers in computer programs, a base of 16 is often used because it's easy to translate from binary. As there are only 10 symbols for numbers (0–9), the values for 10–16 are represented by the letters A–F.

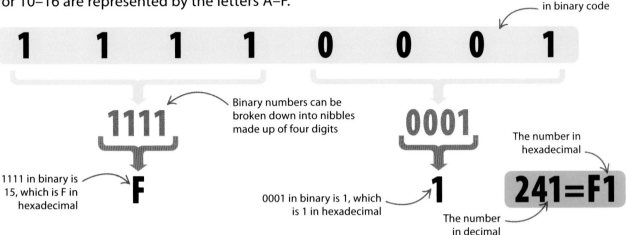

The number 241 in binary code

Binary numbers can be broken down into nibbles made up of four digits

1111 in binary is 15, which is F in hexadecimal

0001 in binary is 1, which is 1 in hexadecimal

The number in hexadecimal

The number in decimal

241=F1

▽ **Comparing base systems**

Using this table, you can see that expressing numbers in hexadecimal gives the most information with the fewest digits.

DIFFERENT BASES		
Decimal	**Binary**	**Hexadecimal**
0	0 0 0 0	0
1	0 0 0 1	1
2	0 0 1 0	2
3	0 0 1 1	3
4	0 1 0 0	4
5	0 1 0 1	5
6	0 1 1 0	6
7	0 1 1 1	7
8	1 0 0 0	8
9	1 0 0 1	9
10	1 0 1 0	A
11	1 0 1 1	B
12	1 1 0 0	C
13	1 1 0 1	D
14	1 1 1 0	E
15	1 1 1 1	F

REMEMBER

Bits, nibbles, and bytes

A binary digit is known as a "bit", and is the smallest unit of memory in computing. Bits are combined to make "nibbles" and "bytes". A kilobit is 1024 bits. A megabit is 1024 kilobits.

1

Bits: Each bit is a single binary digit – a 1 or 0.

1001

Nibbles: Four bits make up a nibble – enough for one hexadecimal digit.

10110010

Bytes: Eight bits, or two hexadecimal digits, make up a byte. This gives us a range of values from 0 to 255 (00 to FF).

Symbols and codes

Computers use binary code to translate numbers into electrical signals. But how would a computer use binary to store the words and characters on this page?

SEE ALSO

❬ **180–181** Inside a computer

❬ **182–183** Binary and bases

ASCII

The first computers each stored characters in their own unique way. This worked fine until data needed to be moved between computers. At this point, a common system was chosen, called the American Standard Code for Information Interchange (ASCII, pronounced "askey").

▷ **ASCII table**
In ASCII, a decimal number value is given to each character in the upper- and lower case alphabets. Numbers are also assigned to punctuation and other characters, such as a space.

▷ **ASCII in binary**
As each character has a number, that number then needs to be converted to binary to be stored in a computer.

R = 82 = 1010010

r = 114 = 1110010

▽ **ASCII in Python**
You can convert between ASCII and binary in most languages, including Python.

This command prints the character, the ASCII value, and the binary value for each letter in the name "Sam"

```
>>> name = "Sam"
>>> for c in name:
        print(c, ord(c), bin(ord(c)))
```

```
S 83 0b1010011
a 97 0b1100001
m 109 0b1101101
```

Here are the results. The beginning of each binary number is marked "0b"

ASCII						
32	SPACE	64	@	96	`
33	!	65	A	97	a
34	"	66	B	98	b
35	#	67	C	99	c
36	$	68	D	100	d
37	%	69	E	101	e
38	&	70	F	102	f
39	'	71	G	103	g
40	(	72	H	104	h
41	)	73	I	105	i
42	*	74	J	106	j
43	+	75	K	107	k
44	,	76	L	108	l
45	-	77	M	109	m
46	.	78	N	110	n
47	/	79	O	111	o
48	0	80	P	112	p
49	1	81	Q	113	q
50	2	82	R	114	r
51	3	83	S	115	s
52	4	84	T	116	t
53	5	85	U	117	u
54	6	86	V	118	v
55	7	87	W	119	w
56	8	88	X	120	x
57	9	89	Y	121	y
58	:	90	Z	122	z
59	;	91	[	123	{
60	<	92	\	124	|
61	=	93	]	125	}
62	>	94	^	126	~
63	?	95	_	127	DELETE

Unicode

As computers across the world began to share data, the limits of ASCII began to show. Thousands of characters used in hundreds of languages had to be represented, so a universal standard called Unicode was agreed on.

Unicode has over 110,000 characters!

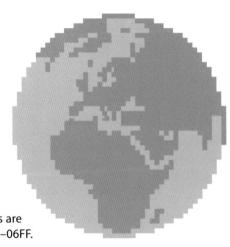

▷ **International code**
Unicode represents all the languages of the world. For example, the Arabic characters are represented in the range 0600–06FF.

▷ **Unicode characters**
Unicode characters are represented by their hexadecimal value, which appears as a series of letters and numbers (see pp.182–183). Each character has its own code. More characters are added all the time, and there are some unusual ones, such as a mini umbrella.

2602

2EC6

08A2

0036

0974

004D

2702

A147

REMEMBER

Hexadecimals

Hexadecimal numbers have a base of 16. Ordinary decimal numbers are used for 0 to 9, and the values 10–15 are represented by the letters A to F. Each hexadecimal number has an equivalent binary value.

The Unicode value of ë as hexadecimal

The same value as binary

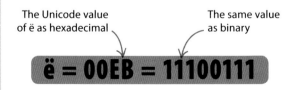

ë = 00EB = 11100111

▽ **Unicode in Python**
Unicode can be used to display special characters in Python. Simply type a string containing a Unicode character code.

Putting "\u" before the hexadecimal code tells the computer this is Unicode

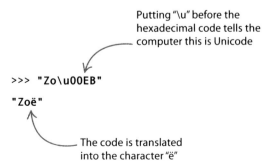

```
>>> "Zo\u00EB"
"Zoë"
```

The code is translated into the character "ë"

Logic gates

As well as to understand numbers and letters, computers can also use electrical signals to make decisions using devices called "logic gates". There are four main types of logic gates: "AND", "NOT", "OR", and "EXCLUSIVE OR".

SEE ALSO

❰ **180–181** Inside a computer

❰ **182–183** Binary and bases

AND gate

Gates use one or more input signals to produce an output signal, based on a simple rule. AND gates switch their output signal "on" (1) only when both input signals are "on" (1 *and* 1).

△ **Inputs 1 and 1 = output 1**
Both input signals are "on", so the AND gate produces an "on" output signal.

△ **Inputs 1 and 0 = output 0**
If one input is "on" but the other is "off", the output signal is "off".

△ **Inputs 0 and 0 = output 0**
An AND gate produces an "off" output signal if both input signals are "off".

NOT gate

These gates "flip" any input to its opposite. "On" input becomes "off" output, and "off" input turns to "on" output. NOT gates are also known as "inverters".

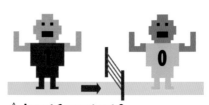

△ **Input 1 = output 0**
The NOT gate flips an "on" input to an "off" output, and vice versa.

OR gate

An OR gate produces an "on" output when either one of the inputs is "on", or when both are "on".

△ **Inputs 1 and 1 = output 1**
Two "on" inputs produce an "on" output.

△ **Inputs 1 and 0 = output 1**
One "on" and one "off" input still produce an "on" output.

△ **Inputs 0 and 0 = output 0**
Only two "off" inputs produce an "off" output from an OR gate.

EXCLUSIVE OR gate

This type of gate only gives an "on" output when one input is "on" and the other is "off". Two "on" or two "off" inputs will produce an "off" output. Gates like this are also known as "XOR" gates.

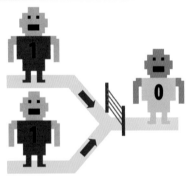

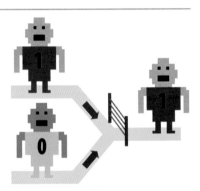

△ **Inputs 1 and 1 = output 0**
Two "on" inputs produce an "off" output.

△ **Inputs 1 and 0 = output 1**
The output is only "on" when the inputs are different.

▪▪ EXPERT TIPS

Building computer circuits

By combining these four basic logic gates, you can create circuits to perform a whole range of advanced functions. For example, by linking an AND gate to an XOR gate, you create a circuit that can add two binary digits (bits) together. By linking two OR gates

with two NOT gates in a loop, you can create a circuit that will store a bit of data (a single 1 or 0). Even the most powerful computers are based on billions of tiny logic circuits.

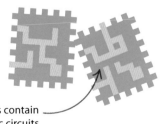

Computer chips contain many logic circuits

Processors and memory

Inside a computer are many types of electronic chips. Most importantly, the processor chip runs programs and memory chips store data for instant access.

SEE ALSO

‹ **180–181** Inside a computer

‹ **186–187** Logic gates

The processor

Processors are a collection of very small and complex circuits, printed on a glass-like material called silicon. Small switches called transistors are combined to form simple logic gates, which are further combined to form complex circuits. These circuits run all the programs on your computer.

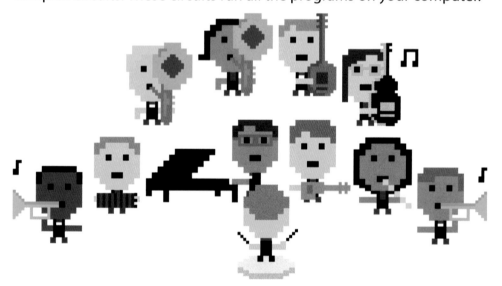

◁ **Circuits in a processor**
The circuits are kept synchronized by a clock pulse, just like an orchestra is kept in time by a conductor.

Machine code

Processors only understand a set of program instructions called "machine code". These simple instructions for operations like adding, subtracting, and storing data are combined to create complex programs.

▷ **Understanding machine code**
Machine code is just numbers, so coders use programming languages like Python that get converted into machine code.

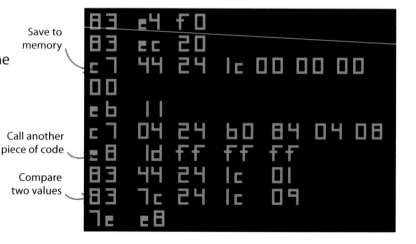

Save to memory

Call another piece of code

Compare two values

```
83  e4  f0
83  ec  20
c7  44  24  1c  00  00  00
00
eb  11
c7  04  24  b0  84  04  08
e8  1d  ff  ff  ff
83  44  24  1c  01
83  7c  24  1c  09
7e  e8
```

Memory

Like processors, memory chips are printed on silicon. A few logic gates are combined to create a "latch circuit". Each latch stores one bit (the smallest unit of data with a binary value of either 1 or 0), and many latches are combined to create megabytes and gigabytes of storage.

Memory is made up of repeated identical blocks of circuit

Every item of data has a number (called an "address") so it can be found quickly

◁ **Programs and data**
Programs constantly read, write, and update the data stored in the memory.

Each block of memory can store millions or billions of bits of data

Processing information

The processor and memory, when combined with input and output devices, give you everything you need for a computer. In a game program, for example, the user inputs position data by clicking the mouse, the processor does the calculations, reads and writes memory, and then produces output in the form of making the character jump on the screen.

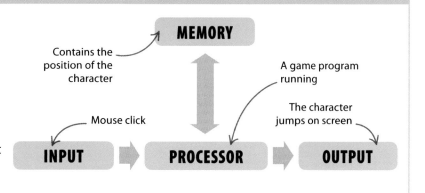

Contains the position of the character

MEMORY

A game program running

The character jumps on screen

Mouse click

INPUT ➡ **PROCESSOR** ➡ **OUTPUT**

Essential programs

There are a few programs that every computer needs in order to work. Some of the most important programs are operating systems, compilers, and interpreters.

SEE ALSO

‹ 180–181 Inside a computer

‹ 182–183 Binary and bases

‹ 188–189 Processors and memory

Operating system

The operating system (OS) is the manager of the computer's resources. It controls which programs are allowed to run, how long they run for, and which parts of the computer they use while running. The OS also provides interfaces, such as file browsers, to let a user interact with the computer. Common operating systems include Microsoft Windows and macOS.

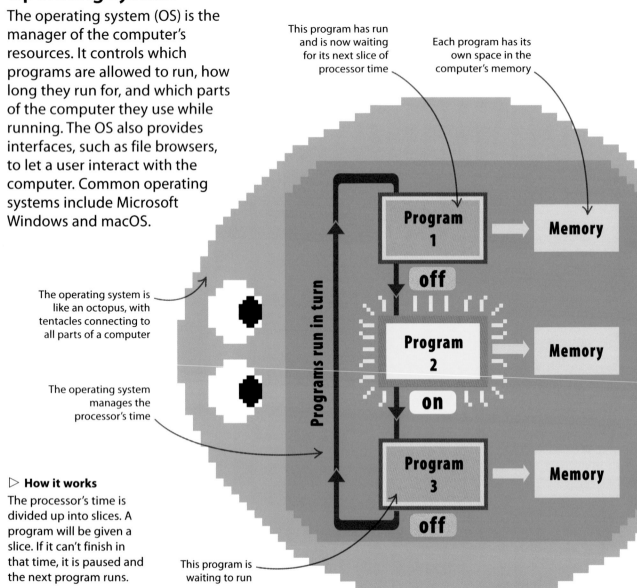

This program has run and is now waiting for its next slice of processor time

Each program has its own space in the computer's memory

The operating system is like an octopus, with tentacles connecting to all parts of a computer

The operating system manages the processor's time

Programs run in turn

Program 1 → Memory

off

Program 2 → Memory

on

Program 3 → Memory

off

▷ **How it works**
The processor's time is divided up into slices. A program will be given a slice. If it can't finish in that time, it is paused and the next program runs.

This program is waiting to run

Compilers and interpreters

The languages you write programs with, such as Python, are known as "high-level languages". Computer processors don't understand these languages, so compilers and interpreters are used to translate them into a low-level language (known as "machine code") that a computer does understand.

Compiler

Program → output → Run compiler → output → Program in machine code

input data → Run program in machine code → output data

△ **Compiler**
Compilers produce translated machine code that can be saved and run later.

Interpreter

▷ **Interpreter**
Interpreters translate the code and execute the program at the same time.

Program → input → Interpreter runs program

input data → Interpreter runs program → output data

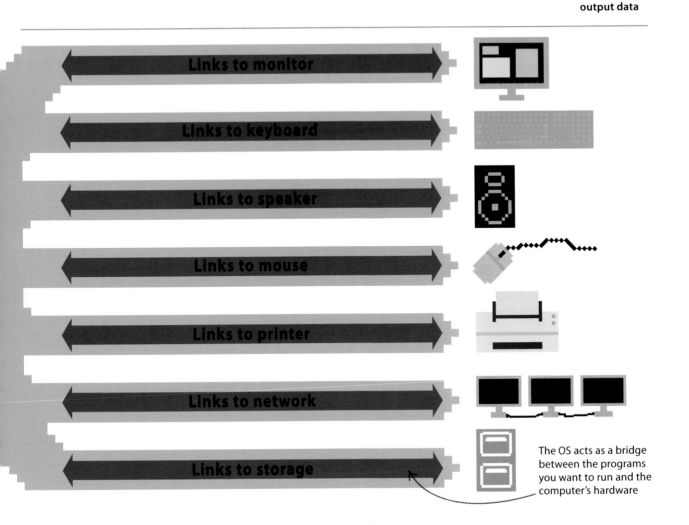

Links to monitor

Links to keyboard

Links to speaker

Links to mouse

Links to printer

Links to network

Links to storage

The OS acts as a bridge between the programs you want to run and the computer's hardware

Storing data in files

A computer's memory doesn't just store numbers and characters. Many more types of data can be stored, including music, pictures, and videos. But how is this data stored? And how can it be found again?

SEE ALSO

❮ **182–183** Binary and bases

❮ **188–189** Processors and memory

❮ **190–191** Essential programs

How is data stored?

When data is saved to be used later, it is put into a file. This file can be given a name that will make it easy to find again. Files can be stored on a hard-drive, memory stick, or even online – so data is safe even when a computer is switched off.

The computer's file system is similar to a paper filing system

EXPERT TIPS

File sizes

Files are essentially collections of data in the form of binary digits (bits). File sizes are measured in the following units:

Bytes (B)

1 B = 8 bits (for example, 10011001)

Kilobytes (KB)

1 KB = 1024 B

Megabytes (MB)

1 MB = 1024 KB = 1,048,576 B

Gigabytes (GB)

1 GB = 1024 MB = 1,073,741,824 B

Terabytes (TB)

1 TB = 1024 GB = 1,099,511,627,776 B

▽ **File information**

There is more to a file than just its contents. File properties tell the system everything it needs to know about a file.

Right-click on a file to see properties such as file type, location, and size

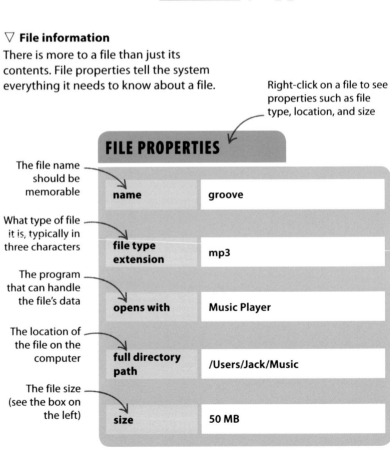

FILE PROPERTIES

The file name should be memorable

| name | groove |

What type of file it is, typically in three characters

| file type extension | mp3 |

The program that can handle the file's data

| opens with | Music Player |

The location of the file on the computer

| full directory path | /Users/Jack/Music |

The file size (see the box on the left)

| size | 50 MB |

Directories

It's easier to find files on a computer system if they are well organized. To help with this, files can be grouped together in "directories", also known as "folders". It's often useful for directories to contain other directories in the form of a directory tree.

▽ **Directory tree**
When directories are placed inside other directories, it creates a structure that resembles an upside-down tree, and just like a tree it has roots and branches (confusingly called "paths").

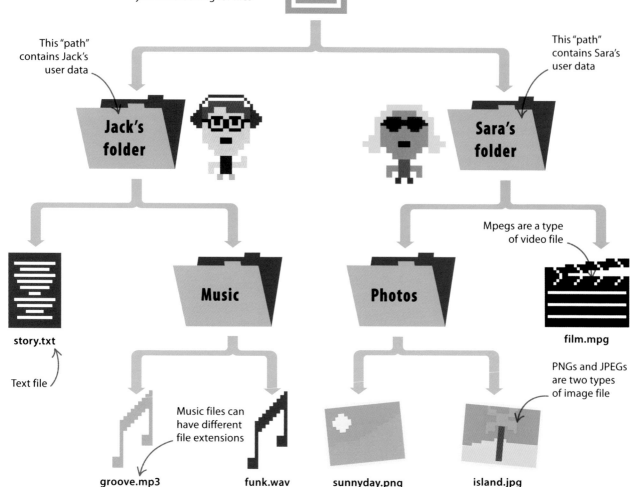

The "root" of the directory tree, where you start looking for files

This "path" contains Jack's user data

This "path" contains Sara's user data

Jack's folder

Sara's folder

Mpegs are a type of video file

Music

Photos

film.mpg

story.txt

Text file

Music files can have different file extensions

PNGs and JPEGs are two types of image file

groove.mp3

funk.wav

sunnyday.png

island.jpg

The internet

The internet is a network of computers all across the world. With so many computers, clever systems are needed to make sure information goes to the right place.

SEE ALSO

❰ **182–183** Binary and bases

❰ **192–193** Storing data in files

IP addresses

Every computer or phone connected to the internet has an address, much like a building. The addresses are called "Internet Protocol (IP) addresses" and each one is made up of a series of numbers.

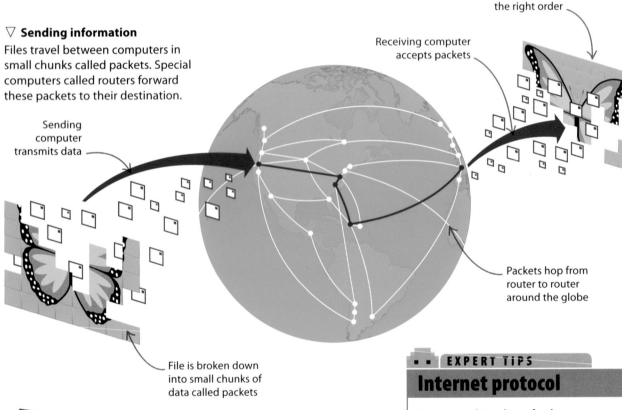

▽ **Sending information**
Files travel between computers in small chunks called packets. Special computers called routers forward these packets to their destination.

Sending computer transmits data

Receiving computer accepts packets

Packets are put back together in the right order

Packets hop from router to router around the globe

File is broken down into small chunks of data called packets

to...
10.150.93.22

from...
62.769.20.57

◁ **Address information**
Every packet of data is labelled with the destination and sender's IP addresses. Domain names like "dk.com" are translated into IP addresses.

· · EXPERT TIPS

Internet protocol

A protocol is a list of rules. "Internet Protocols" are rules for how big packets can be and how they are structured. All internet devices must follow these rules if they want to be able to communicate with each other.

Moving data

Before packets can be sent between devices, they have to be translated into binary signals (ones and zeroes) that can travel over great distances. Every device on the internet has a "network adapter" to perform this task. Different devices send data in different forms.

△ **Electrical signals**
Copper wires carry ones and zeroes as electrical signals of different strengths.

△ **Light**
Special glass fibres, called fibre optic cables, transmit data as pulses of light.

△ **Radio waves**
Different types of radio waves can carry ones and zeroes without using wires.

Ports

Just as you post a letter to a specific person in an apartment building, you may want to send packets to a specific program on a device. Computers use numbers called "ports" as addresses for individual programs. Some common programs have ports specially reserved for them. For example, web browsers always receive packets through port number 80.

▽ **Port numbers**
The numbers used for ports range from 0 to 65535 and are divided into three types: well-known, registered, and private.

A device's IP address is like the street address of a building

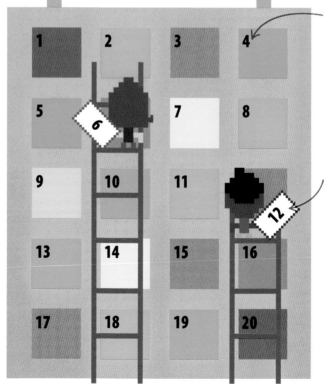

IP 165.193.128.72

A port within a device is like an apartment in a building

Routers deliver packets like postmen to the correct addresses

Sockets

The combination of an IP address and a port is known as a "socket". Sockets let programs send data directly to each other across the internet, which is useful for things such as playing online games.

Programming in
the real world

Computer languages

Thousands of different programming languages have been created. Which one you should use depends on a number of factors, such as the type of program being written and which kind of computer it will run on.

SEE ALSO

Computer **204–205 ›**
games

Making **206–207 ›**
apps

Popular programming languages

Some languages have emerged as the most popular for creating certain types of program on certain types of computer. Here is how to run a simple "Hello World!" program in a few popular programming languages.

```
#include <stdio.h>
main()
{
        printf("Hello World!");
}
```

△ **C**
One of the most popular languages of all time, C is often used for programming hardware.

```
#include <iostream>
int main()
{
        std::cout << "Hello World!" << std::endl;
}
```

△ **C++**
Based on C, but with extra features. Used in programs that need to be fast, such as console games.

```
#import <stdio.h>
int main(void)
{
    printf("Hello World!");
}
```

△ **Objective-C**
Based on C, with some extra features. It has become popular due to its use on Apple's Mac and iOS devices.

```
alert('Hello World!');
```

△ **JavaScript**
Used to create programs that run on web browsers, such as simple games and email websites.

```
class HelloWorldApp {
    public static void main(String[] args) {
        System.out.println("Hello World!");
    }
}
```

△ **Java**
A very versatile language that can run on most computers. It's often used for coding on the Android operating system.

```
<?php
echo "Hello World!";
?>
```

◁ **PHP**
Mostly used for creating interactive websites, PHP runs on the web servers that host websites.

Languages from the past

Many languages that were famous twenty or thirty years ago have fallen in popularity, despite still being used in some very important systems. These languages are often seen as difficult to code by modern standards.

BASIC
Designed in 1964 at Dartmouth College, in the USA, BASIC was very popular when home computers first became available.

Fortran
Designed in 1954 at IBM, a technology firm, Fortran is mainly used for calculations on large computers. It is still being used in weather forecasting.

COBOL
Designed in 1959 by a committee of experts, COBOL is still being used in many business and banking programs.

▪▪ REAL WORLD
Millennium bug

Many programs in older languages like COBOL used two digits to represent a year (such as 99 for 1999). The "millennium bug" was predicted to cause problems in 2000 when these dates rolled over into the new millennium as 00.

Computers all over the world had to be updated to stop the millennium bug

Weird languages

Among the thousands of languages are a few that have been created for very specific and strange purposes.

('&%:9]!~}|z2Vxwv-,POqponl$Hjig%eB@@>a=<M:9[p6tsl1TS/QIOj)L(I&%$""Z~AA@UZ=RvttT`R5P3m0LEDh,T*?(b&`$#87[]{W

△ **Malbolge**
The Malbolge language was designed to be impossible to program. The first working code did not emerge until two years after its release, and was written by another program.

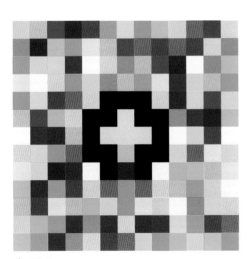

△ **Piet**
Programs created in Piet code look like abstract art. The "Hello World!" program is shown above.

△ **Chef**
A program written in Chef is meant to resemble a cooking recipe. However, in practice, the programs rarely produce useful cooking instructions.

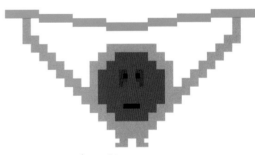

△ **Ook!**
Designed to be used by orangutans, Ook! has only three elements: "Ook", "Ook!", and "Ook?" These can be combined to create six commands, such as "Ook! Ook" and "Ook? Ook!"

Coding stars

Computing is driven forward every day by millions of programmers all around the world, but every now and then someone special comes along and takes a massive leap. Here are a few of the most famous coders.

SEE ALSO

❮ 18–19 Becoming a coder

Computer 204–205 ❯ games

Ada Lovelace

Nationality: British

Dates: 1815–52

Famous for: Ada Lovelace is considered to be the first computer programmer. In 1843, she published the first program for Charles Babbage's Analytical Engine (a proposed early computer). She also suggested methods for representing characters as numbers.

Alan Turing

Nationality: British

Dates: 1912–54

Famous for: Mathematician Alan Turing is known as the father of computer science. He's also famous for his ground-breaking work cracking secret German codes during World War II for the British.

Grace Hopper

Nationality: American

Dates: 1906–92

Famous for: Grace Hopper created the first programming language compiler (which transforms human readable programs into machine code). As well as being a computer scientist she was a Rear Admiral in the US Navy!

Bill Gates and Paul Allen

Nationality: American

Dates: Gates 1955–present, Allen 1953–present

Famous for: Bill Gates and Paul Allen founded Microsoft together in the 1970s. They invented some of the most popular programs ever, such as Microsoft Windows and Office.

Gunpei Yokoi and Shigeru Miyamoto

Nationality: Japanese

Dates: Yokoi 1941–97, Miyamoto 1952–present

Famous for: Yokoi and Miyamoto worked for Nintendo, the gaming company. Yokoi invented the Game Boy, while Miyamoto made successful games such as Super Mario.

Tim Berners-Lee

Nationality: British

Dates: 1955–present

Famous for: While working at CERN (a famous scientific research centre in Switzerland), Tim Berners-Lee invented the world wide web, and made it free for everyone. He was knighted by Queen Elizabeth II in 2004.

Larry Page and Sergei Brin

Nationality: American

Dates: Both 1973–present

Famous for: In 1996, Page and Brin began work on what would become the Google search engine. Their effective search method revolutionized the internet.

Mark Zuckerberg

Nationality: American

Dates: 1984–present

Famous for: Zuckerberg launched Facebook from his college room in 2004. Facebook has since become a billion-dollar company, and made Zuckerberg one of the wealthiest people alive.

Open Source Movement

Nationality: All

Dates: Late 1970s–present

Famous for: The open source movement is a collection of programmers around the world who believe software should be free and available to all. The movement has been responsible for many significant pieces of software, such as the GNU/Linux operating system and Wikipedia, the online encyclopedia.

Busy programs

Computers and programs have become an invisible part of daily life. Every day, people benefit from very complex computer programs that have been written to solve incredibly tough problems.

SEE ALSO

❮ 180–181 Inside a computer

❮ 192–193 Storing data in files

Compressing files

Almost every type of file that is sent over the internet is compressed (squeezed) in some way. When a file is compressed, data that isn't needed is identified and thrown away, leaving only the useful information.

◁ **Squeezing data**
Compressing a file is like squeezing a jack-in-the-box to make it fit into a smaller space.

· · ·   REAL WORLD

Music files

Without music compression programs, you could only fit a few songs on your music player. By compressing audio files, the average smartphone can now hold thousands of songs.

Secret codes

When you log in to a website, buy something, or send a message across the internet, smart programs scramble your secret data so that anyone who intercepts it won't be able to understand it. Global banking systems rely on these advanced programs capable of hiding secret information.

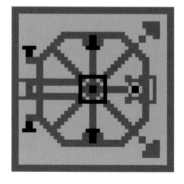

◁ **Cryptography**
Cryptography is the study of codes. Complex mathematical codes scramble and unscramble personal data to keep it safe from thieves.

Artificial Intelligence

Intelligent programs do more than just make computer games fun. Artificial Intelligence (AI) is being used to provide better healthcare, as well as helping robots operate in places too dangerous for humans to go, such as warzones and areas destroyed by natural disasters.

△ **Medicine**
Systems are able to analyse a huge database of medical information and combine it with details from the patient to suggest a diagnosis.

△ **Bomb disposal**
Many soldiers' lives can be saved by using an intelligent robot to safely dispose of a bomb in an area that has been cleared of people.

Supercomputers

Supercomputers – used by high-tech organizations such as NASA – combine the power of thousands of computer processors that share data and communicate quickly. The result is a computer that can perform millions of calculations per second.

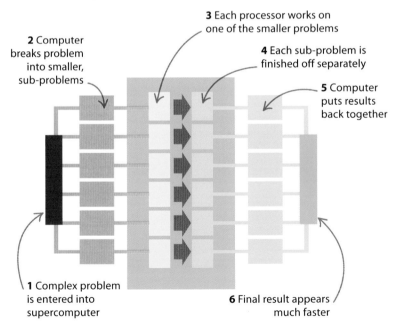

2 Computer breaks problem into smaller, sub-problems

3 Each processor works on one of the smaller problems

4 Each sub-problem is finished off separately

5 Computer puts results back together

1 Complex problem is entered into supercomputer

6 Final result appears much faster

△ **How it works**
Problems are broken into smaller problems that are all worked on separately at the same time by different processors. The results are then combined together to give the answer.

Weather forecast

Weather patterns are very unpredictable. Supercomputers crunch the huge amounts of data needed to accurately predict what will happen. Each processor in the supercomputer calculates the weather for a small part of the map. All the results are then combined to produce the whole forecast.

Computer games

What does it take to make a modern video game? All computer games are a different mix of the same ingredients. Great games are usually made by teams of software developers – not just programmers.

SEE ALSO

❮ **200–201** Coding stars

Making **206–207** ❯ apps

Who makes computer games?

Even simple games on your mobile phone might be made by large teams of people. For a game to be popular and successful, attention to detail needs to be given to every area during its development, which involves many people with lots of different skills.

△ **Graphic designer**
All of the levels and characters need to look good. The graphic designers define the structure and appearance of everything in the game.

△ **Coder**
Programmers write the code that will make the game work, but they can only do this with input from the rest of the team.

◁ **Level designer**
The architects of the game's virtual world, level designers create settings and levels that are fun to play.

▷ **Tester**
Playing games all day may seem like a great job, but testers often play the same level over and over again to check for bugs.

△ **Scriptwriter**
Modern games have interesting plots just like great books and films. Scriptwriters develop all the characters and stories for the game.

◁ **Sound designer**
Just like a good movie, a great game needs to have quality music and sound effects to set the mood.

LINGO

Consoles

A console is a special type of computer that is well suited to running games. Consoles, such as the PS4 and Xbox One, often have advanced graphics and sound processors capable of running many things at once, making more realistic games possible.

Game ingredients

The most common ingredients in games are often combined into a "game engine". Engines provide an easy-to-use base so that new games can be developed quickly.

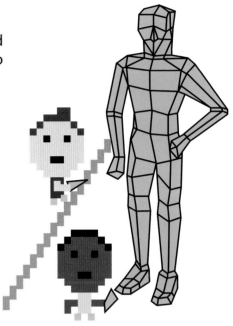

▷ **Story and game logic**

All games must have a good story and some sort of goal to aim for, such as saving the princess. Well-designed game logic keeps players interested.

◁ **Game physics**

In a virtual world, the rules of the real world, such as gravity and collisions, must be re-created to make the game more believable.

△ **Graphics**

As games become more realistic their graphics must become more complex. Body movements, smoke, and water are particularly hard to get right.

▷ **Controls**

Familiar controls that make sense to the player help to make a great game. Good control design makes the player forget that they are using a controller.

▷ **Sound**

All of the words spoken in the game must be recorded, as well as the background music and the sound effects that change throughout the game.

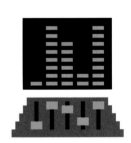

Open the pod bay doors, Hal

I'm sorry Dave, I'm afraid I can't do that

△ **Artificial intelligence**

Human players often play alongside or against computer-controlled players. Artificial intelligence programming allows these characters to respond realistically.

Serious games

Games are being used for more than just fun. Pilots, surgeons, and soldiers are just some of the professionals who use games at work for training purposes. Some businesses even use strategy games to improve their employees' planning skills.

Making apps

Mobile phones have opened up a world of possibilities for coders. With a computer in everyone's pocket, mobile apps can use new inputs, such as location-finding and motion-sensing, to give users a better experience.

SEE ALSO

❰ **190-191** Essential programs

❰ **198–199** Computer languages

❰ **204–205** Computer games

What is an app?

"App" (short for "application") is a word that describes programs that run on mobile devices, including smartphones, tablets, and even wearable technology such as watches. There are many different categories of apps that do different things.

◁ **Social network**
Social apps can allow people to connect with friends, whether they are nearby or far away, to share thoughts, pictures, music, and videos.

◁ **Games**
All sorts of games are available on mobile devices, from simple puzzle games to fast-paced action adventures.

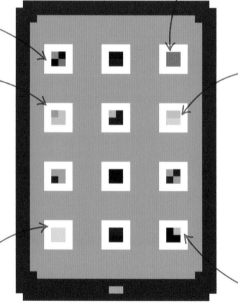

△ **Travel**
Travel apps use your location combined with other users' reviews to provide recommendations for restaurants, hotels, and activities.

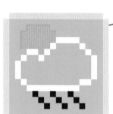

△ **Weather**
Mobile apps use your location to provide accurate weather forecasts, and also allow you to check the weather around the world.

◁ **Sport**
People use apps to track their fitness when running or cycling, and can also keep up to date on the latest sports scores while on the go.

△ **Education**
Educational apps are great for learning. Young children can learn to count and spell, and older people can learn a new language.

How to build an app

There are many questions to answer before building an app. What will it do? What devices will it run on? How will the user interact with it? Once these questions are answered, building an app is a step-by-step process.

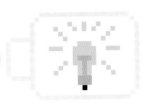

Mac

Android

Windows

1 Have an idea
Any idea for a new app must be well suited to mobile devices. It might be a completely new idea, or it could just be an improvement on an already existing idea to make a better version.

2 Which operating system?
Will the app target a certain type of mobile device? Coders can often use tools that let them write their application once and then adapt it for different operating systems.

3 Learn to make apps
Whichever platform the app will run on, a coder needs to learn the language and other skills needed to build a good app. Online tutorials and local coding clubs can help.

4 Create the program
Good apps take time to make. A basic version might be working in weeks, but for an app to be really successful, it will need to be developed for a few months before its release.

5 Test it
Users will quickly get rid of an app if it contains bugs. Putting in tests as part of the code, and getting friends and family to try out the app can help to clean up any errors before the app is released.

Programming for the internet

SEE ALSO

❮ **198–199** Computer languages

Using **210–211** ❯ JavaScript

Websites are built using coding languages that work just like Python. One of the most important of these is JavaScript, which makes websites interactive.

How a web page works

Most web pages are built using several different languages. An email website, for instance, is made with CSS, HTML, and JavaScript. The JavaScript code makes the site respond instantly to mouse clicks without having to reload the page.

◁ **CSS**
The language CSS (Cascading Style Sheets) controls the colours, fonts, and layout of the page.

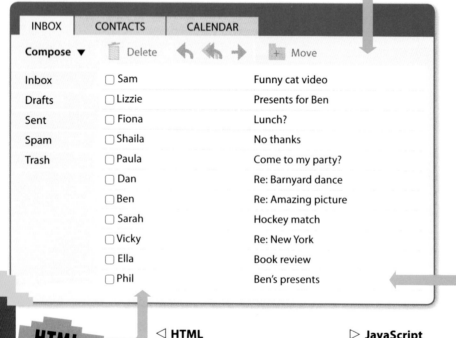

INBOX	CONTACTS	CALENDAR	

Compose ▼   🗑 Delete   ↩ ↩ →   ➕ Move

Inbox	☐ Sam	Funny cat video
Drafts	☐ Lizzie	Presents for Ben
Sent	☐ Fiona	Lunch?
Spam	☐ Shaila	No thanks
Trash	☐ Paula	Come to my party?
	☐ Dan	Re: Barnyard dance
	☐ Ben	Re: Amazing picture
	☐ Sarah	Hockey match
	☐ Vicky	Re: New York
	☐ Ella	Book review
	☐ Phil	Ben's presents

◁ **HTML**
HTML (HyperText Markup Language) builds the basic structure of the page, with different sections that contain text or images.

▷ **JavaScript**
JavaScript controls how the page changes when you use it. Click on an email, for instance, and JavaScript makes a message open up.

HTML

When you open a website, your internet browser downloads an HTML file and runs the code to turn it into a web page. To see how it works, type the code here into an IDLE code window (see pp.92–93) and save it as a file with the ending ".html". Double click the file and it will launch a browser window saying "Hello World!"

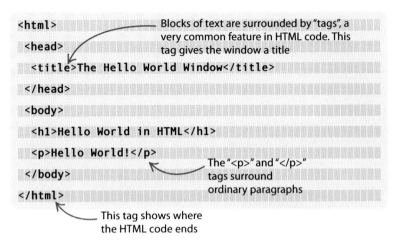

```
<html>
  <head>
   <title>The Hello World Window</title>
  </head>
  <body>
   <h1>Hello World in HTML</h1>
   <p>Hello World!</p>
  </body>
</html>
```

Blocks of text are surrounded by "tags", a very common feature in HTML code. This tag gives the window a title

The "<p>" and "</p>" tags surround ordinary paragraphs

This tag shows where the HTML code ends

Trying JavaScript

It's easy to experiment with JavaScript as all modern web browsers can understand it. JavaScript code is usually placed within HTML code, so the example below uses two coding languages at once. The JavaScript section is surrounded by "<script>" tags.

1 **Write some JavaScript**
Open a new IDLE code window and type out the code below. Check the code very carefully. If there are any errors, you'll just see a blank page.

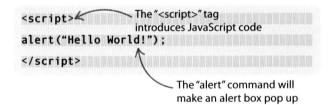

```
<script>
alert("Hello World!");
</script>
```

The "<script>" tag introduces JavaScript code

The "alert" command will make an alert box pop up

2 **Save your file**
Save the file and enter a filename such as "test.html" so the code is saved as an HTML file and not a Python file. Then double click the file to test it.

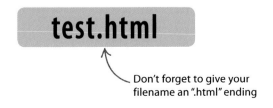

Don't forget to give your filename an ".html" ending

EXPERT TIPS

Games in JavaScript

JavaScript is so good at creating interactive features that it can be used to make games – from simple puzzles to fast-paced racing games. These will work in any modern web browser, so there's no need to install the game first. JavaScript is also used to create web apps such as webmail or interactive calendars.

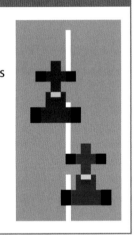

3 **Pop-up appears**
The browser will open and an interactive alert box will pop up with the greeting "Hello World!" Click "OK" to dismiss the box.

JavaScript creates interactive features such as buttons

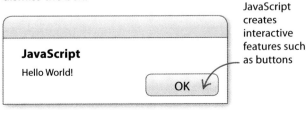

JavaScript

Hello World!

OK

Using JavaScript

JavaScript is great for creating mini programs that run inside HTML, bringing websites to life and allowing users to interact with them. Although it works like Python, JavaScript code is more concise and trickier to learn.

SEE ALSO

‹ 162–163 Reacting to events

‹ 122–123 Loops in Python

‹ 208–209 Programming for the internet

Getting input

As with Python, you can use JavaScript to ask the user for information. JavaScript can do this with a pop-up box. The following program prompts the user to enter their name and responds with a greeting.

This line creates a pop-up box and stores the text the user types into it

1 **Use a prompt**
This short script stores the user's name in a variable. Type the code into the IDLE code window and remember to save it with a ".html" filename.

```
<script>
var name = prompt("Please enter your name");
var greeting = "Hello " + name + "!";
document.write(greeting);
</script>
```

The text in quotes appears in the box

JavaScript lines always end with a semicolon

The "</script>" tag shows where the JavaScript ends

This line displays the greeting

2 **Question appears**
Double-click the HTML file to launch a browser window. Enter your name in the box and click "OK" to see the greeting.

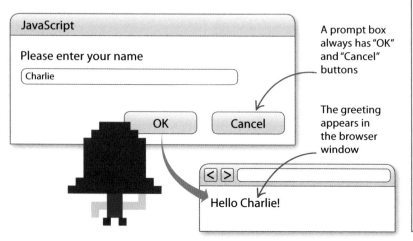

JavaScript

Please enter your name

Charlie

OK   Cancel

A prompt box always has "OK" and "Cancel" buttons

The greeting appears in the browser window

Hello Charlie!

• • **EXPERT TIPS**

Type carefully

When working with JavaScript, be careful to check that you've typed out the code correctly. If there's an error, the browser will simply ignore the whole block of JavaScript and will create a blank window, without any error message saying what went wrong. If that happens, check the code again carefully.

Events

An event is any action that a program can detect, such as a mouse click or a keystroke. The section of code that reacts to an event is called an "event handler". Event handlers are used a lot in JavaScript and can trigger many different functions, making web pages fun and interactive.

1 Type the code
In this example, an event (clicking a button) triggers a simple function (a tongue-twister appears). Type the code in an IDLE code window and save the file with a ".html" ending.

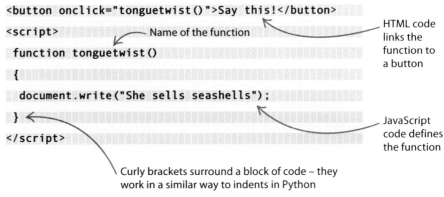

```
<button onclick="tonguetwist()">Say this!</button>
<script>
function tonguetwist()
{
  document.write("She sells seashells");
}
</script>
```

Name of the function

HTML code links the function to a button

JavaScript code defines the function

Curly brackets surround a block of code – they work in a similar way to indents in Python

2 Run the program
Double-click the file to launch the program in a browser window.

Click the button

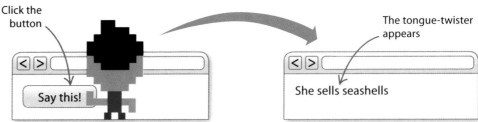

The tongue-twister appears

Say this!

She sells seashells

Loops in JavaScript

A loop is a section of code that repeats. Using loops is much quicker and easier than typing out the same line of code over and over again.

1 Loop code
Like Python, JavaScript uses "for" to set up a loop. The repeated lines of code are enclosed in curly brackets. This loop creates a simple counter that increases by one each time it repeats.

The "<script>" tag introduces the JavaScript code

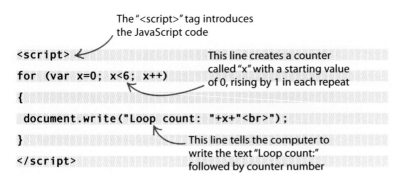

```
<script>
for (var x=0; x<6; x++)
{
document.write("Loop count: "+x+"<br>");
}
</script>
```

This line creates a counter called "x" with a starting value of 0, rising by 1 in each repeat

This line tells the computer to write the text "Loop count:" followed by counter number

2 Loop output
Save the code as a ".html" file and run it. The loop keeps repeating as long as "x" is less than 6 ("x<6" in the code). To increase the number of repeats, use a higher number after the "<" symbol.

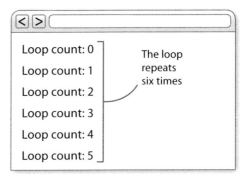

Loop count: 0
Loop count: 1
Loop count: 2
Loop count: 3
Loop count: 4
Loop count: 5

The loop repeats six times

Bad programs

Not all programs are fun games or useful apps. Some programs are designed to steal your data or damage your computer. They will often seem harmless, and you might not realize that you have been a victim.

SEE ALSO

‹ **194–195** The internet

‹ **202–203** Busy programs

Malware

Programs that do things without your knowledge or permission are known as "malware". Unauthorized access to a computer is a crime, but there are many different types of programs that still try to sneak on to your computer.

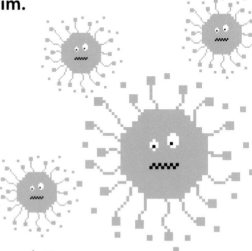

▷ **Worm**

A worm is a type of malware that crawls around a network from computer to computer. Worms can clog up networks, slowing them down – the first worm brought the internet to a virtual standstill in 1988.

△ **Virus**

Just like a virus in the human body, this malware copies itself over and over again. They are usually spread through emails, USB sticks, or other methods of transferring files between computers.

△ **Trojan**

Malware that pretends to be a harmless program is known as a "trojan". The word comes from an ancient war, in which the Greeks gifted the Trojans a giant wooden horse. But the horse had soldiers hidden inside, which helped the Greeks win the war.

■ ■ **REAL WORLD**

Famous worm

On 5 May 2000, internet users in the Philippines received emails with the subject "ILOVEYOU". An attachment appeared to be a love letter, but was actually a piece of malware that corrupted files.

◁ **ILOVEYOU**

The worm quickly spread to computers around the world. It is estimated to have cost more than $20 billion to fix the damage it caused.

What malware does

Viruses, worms, and trojans are all types of malware that want to get into your machine, but what do they do once they have infected their target? They might delete or corrupt files, steal passwords, or seek to control your machine for some larger purpose as part of an organized "zombie botnet".

▷ **Zombie botnets**
Botnets are collections of infected computers that can be used to send spam emails, or flood a target website with traffic to bring it crashing down.

Good software to the rescue

Thankfully, people aren't defenceless in the fight against malware. Anti-malware software has become big business, with many providers competing to provide the best protection. Two well-known examples are firewalls and antivirus programs.

△ **Antivirus programs**
Antivirus software tries to detect malware. It identifies bad programs by scanning files and comparing their contents with a database of suspicious code.

△ **Firewalls**
Firewalls aim to prevent malware and dangerous network traffic from reaching your computer. They scan all incoming data from the internet.

 LINGO

Hackers

Coders that study and write malware are known as "hackers". Those who write malware to commit crimes are known as "black-hat" hackers, and those who write programs to try to protect against malware are known as "white-hat" hackers.

White-hat hacker   **Black-hat hacker**

Mini computers

Computers don't have to be big or expensive. A whole range of small and cheap computers are available. Because of their small size and cost, these computers are being used in lots of new and exciting ways.

SEE ALSO

❮ **180–181** Inside a computer

❮ **202–203** Busy programs

Raspberry Pi

The Pi is a credit-card-sized computer, created to teach the basics of how computers work. For its size it is impressively powerful, with the ability to run similar programs to a modern PC.

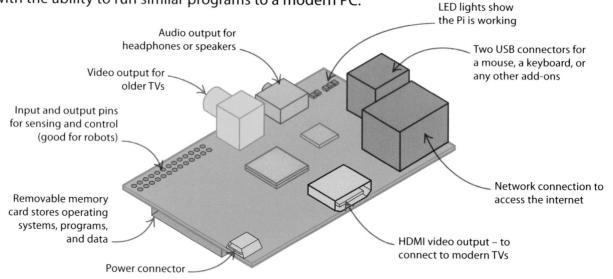

LED lights show the Pi is working

Audio output for headphones or speakers

Two USB connectors for a mouse, a keyboard, or any other add-ons

Video output for older TVs

Input and output pins for sensing and control (good for robots)

Removable memory card stores operating systems, programs, and data

Network connection to access the internet

Power connector

HDMI video output – to connect to modern TVs

Arduino

The Arduino is cheaper than the Pi, but less powerful. It is often used as a low-cost and simple way to build custom electronic or robotic machines.

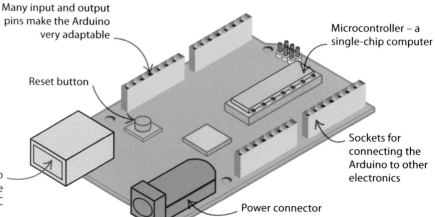

Many input and output pins make the Arduino very adaptable

Microcontroller – a single-chip computer

Reset button

USB connector used to load programs on to the Arduino from a PC

Sockets for connecting the Arduino to other electronics

Power connector

Using mini computers

There are endless useful things a mini computer can do thanks to its many connection options. Here are just a few suggestions.

△ **Computer**
Connect a keyboard, mouse, and monitor for a fully working desktop computer.

△ **Audio output**
Connect a set of speakers and then send music to them over the network.

△ **Mobile phones**
Connect the computer to the internet using a mobile phone.

△ **Gadgets**
Connect LED lights and other simple electronics to make robots or gadgets.

△ **Television**
Connect a TV and use it as a media centre to show all of your movies and pictures.

△ **Camera**
Connect a basic camera to your mini computer to create your own webcam.

△ **USB**
Connect a USB hard drive and share your files over your network.

△ **SD card**
Change the programs on your mini computer just by swapping SD cards.

REAL WORLD

Home-built robots

With their small size, cost, and weight, mini computers are being used more and more to build different types of robot. For example:

Weather balloons that record weather conditions in the atmosphere.

Mini vehicles that can sense obstacles using sonar like a bat.

Robotic arms that pick up and move different objects.

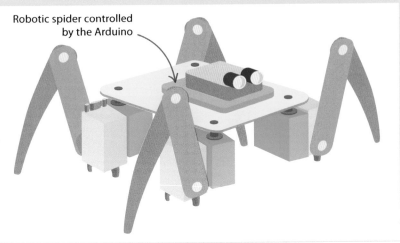

Robotic spider controlled by the Arduino

Become a master programmer

SEE ALSO

❮ **176–177** What next?

❮ **214–215** Mini computers

The secret to becoming a master programmer is to have fun. As long as you're enjoying yourself, there's no limit to how skilled you can become at coding, whether as a hobby or a lifelong career.

Ways to become a better programmer

Like skiing, learning the piano, or playing tennis, coding is a skill that you'll get better and better at over time. It can take years to become a true expert, but if you're having fun on the way, it will feel like an effortless journey. Here are a few tips to help you become a master programmer.

△ **Code a lot**
People say practice makes perfect – and it's true. The more code you write, the better you'll get. Keep going and you'll soon be an expert.

◁ **Be nosy**
Read websites and books about programming and try out other people's code. You'll pick up expert tips and tricks that might have taken you years to figure out on your own.

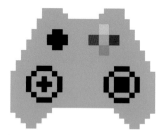

△ **Steal ideas**
If you come across a great program, think how you might code it yourself. Look for clever ideas to use in your own code. All the best programmers copy each other's ideas and try to improve them.

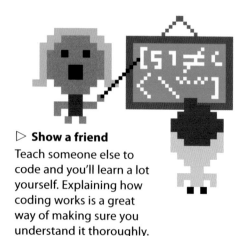

▷ **Show a friend**
Teach someone else to code and you'll learn a lot yourself. Explaining how coding works is a great way of making sure you understand it thoroughly.

▷ **Train your brain**
Your brain is like a muscle – if you exercise it, it will get stronger. Do things that help you think like a programmer. Solve logic puzzles and brainteasers, take up Sudoku, and work on your maths.

▷ Test your code

Test your code by entering crazy values to see what happens. See how well it stands up to errors. Try rewriting it to improve it or try rewriting someone else's – you'll learn all their secret tricks.

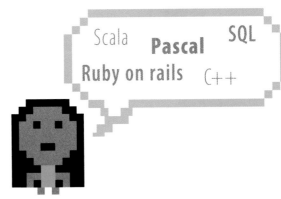

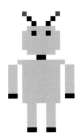

◁ Build a robot army

You can connect your computer to all sorts of programmable devices, from flashing LED lights to robots. It's fun and you'll learn lots as you figure out how to conquer the world.

△ Learn new languages

Become multilingual. Every new programming language you learn will teach you more about the ones you already know (or *thought* you knew). You can download free versions of most languages.

▷ Pull a computer to bits

Take an old computer apart to see how it works (ask permission first!). There aren't many components, so it won't take long to figure out what all the bits are. Best of all, build your own computer and then run your code on it.

▷ Win a prize

When your skills develop, why not enter an online coding contest? There are lots to choose from at all different levels. The toughest are worldwide competitions like Google's Code Jam, but there are easier challenges too.

 REMEMBER

Have fun!

Coding is a lot like trying to solve puzzles. It's challenging and you'll often get stuck. Sometimes it's frustrating. But you'll also have breakthroughs when you solve a problem and feel a buzz of excitement at seeing your code work. The best way to keeping coding fun is to take on challenges that suit you. If a project is too easy you'll get bored; if it's too hard you'll lose interest. Never be afraid to fiddle, tinker, experiment, and break the rules – let your curiosity lead you. But most of all, remember to have fun!

Glossary

algorithm
A set of step-by-step instructions followed when performing a task: for example, by a computer program.

ASCII
"American Standard Code for Information Interchange" – a code used for storing text characters as binary code.

binary code
A way of writing numbers and data that only uses 0s and 1s.

bit
A binary digit – 0 or 1. The smallest unit of digital information.

Boolean expression
A question that has only two possible answers, such as "true" and "false".

branch
A point in a program where two different options are available to choose from.

bug
An error in a program's code that makes it behave in an unexpected way.

byte
A unit of digital information that contains eight bits.

call
To use a function in a program.

compression
A way of making data smaller so that it takes up less storage space.

computer network
A way to link two or more computers together.

container
A part of a program that can be used to store a number of other data items.

data
Information, such as text, symbols, and numerical values.

debug
To look for and correct errors in a program.

debugger
A program that checks other programs for errors in their code.

directory
A place to store files to keep them organized.

encryption
A way of encoding data so that only certain people can read or access it.

event
Something a computer program can react to, such as a key being pressed or the mouse being clicked.

execute
See *run*.

file
A collection of data stored with a name.

float
A number with a decimal point in it.

function
A piece of code that does part of a larger task.

gate
Used by computers to make decisions. Gates use one or more input signals to produce an output signal, based on a rule. For example, "AND" gates produce a positive output only when both input signals are positive. Other gates include "OR" and "NOT".

GPU
A graphics processing unit (GPU) allows images to be displayed on a computer screen.

graphics
Visual elements on a screen that are not text, such as pictures, icons, and symbols.

GUI
The GUI, or graphical user interface, is the name for the buttons and windows that make up the part of the program you can see and interact with.

hacker
A person who breaks into a computer system. "White hat" hackers work for computer security companies and look for problems in order to fix them. "Black hat" hackers break into computer systems to cause harm or to make profit from them.

hardware
The physical parts of a computer that you can see or touch, such as wires, the keyboard, and the display screen.

hexadecimal
A number system based on 16, where the numbers 10 to 15 are represented by the letters A to F.

index number
A number given to an item in a list. In Python, the index number of the first item will be 0, the second item 1, and so on.

input
Data that is entered into a computer: for example, from a microphone, keyboard, or mouse.

integer
Any number that does not contain a decimal point and is not written as a fraction (a whole number).

interface
The means by which the user interacts with software or hardware.

IP address
A series of numbers that makes up a computer's individual address when it is connected to the internet.

library
A collection of functions that can be reused in other projects.

loop
Part of a program that repeats itself (to prevent the need for the same piece of code to be typed out multiple times).

machine code
The basic language understood by computers. Programming languages must be translated into machine code before the processor can read them.

malware
Software that is designed to harm or disrupt a computer. Malware is short for "malicious software".

memory
A computer chip inside a computer that stores data.

module
A section of code that performs a single part of an overall program.

operator
A symbol that performs a specific function: for example, "+" (addition) or "-" (subtraction).

OS
A computer's operating system (OS) provides the basis for other programs to run, and connects them to hardware.

output
Data that is produced by a computer program and viewed by the user.

port
A series of numbers used by a computer as the "address" for a specific program.

processor
A type of electronic chip inside a computer that runs programs.

program
A set of instructions that a computer follows in order to complete a task.

programming language
A language that is used to give instructions to a computer.

random
A function in a computer program that allows unpredictable outcomes. Useful when creating games.

run
The command to make a program start.

server
A computer that stores files accessible via a network.

single-step
A way of making a computer program run one step at a time, to check that each step is working properly.

socket
The combination of an IP address and a port, which lets programs send data directly to each other over the internet.

software
The programs that run on a computer and control how it works.

sprite
A movable object.

statement
The smallest complete instruction a programming language can be broken down into.

string
A series of characters. Strings can contain numbers, letters, or symbols, such as a colon.

syntax
The rules that determine how a program must be structured in order for it to work properly.

Trojan
A piece of malware that pretends to be another piece of software to trick the user.

tuple
A list of items separated by commas and surrounded by brackets.

Unicode
A universal code used by computers to represent thousands of symbols and text characters.

variable
A named place where you can store information that can be changed.

virus
A type of malware that works by multiplying itself to spread between computers.

Index

Page numbers in **bold** refer to main entries.

Acknowledgments

DORLING KINDERSLEY would like to thank: Vicky Short, Mandy Earey, Sandra Perry, and Tannishtha Chakraborty for their design assistance; Olivia Stanford for her editorial assistance; Caroline Hunt and Steph Lewis for proofreading; Helen Peters for the index; and Adam Brackenbury for creative technical support.

DORLING KINDERSLEY INDIA would like to thank: Kanika Mittal for design assistance; Pawan Kumar for pre-production assistance; and Abhijit Dutta and Mark Silas for code testing.

Scratch is developed by the Lifelong Kindergarten Group at MIT Media Lab. See **http://scratch.mit.edu**

Python is copyright © 2001-2013 Python Software Foundation; All Rights Reserved.